Environmental Footprints and Eco-design of Products and Processes

Series editor

Subramanian Senthilkannan Muthu, SgT Group and API,
Hong Kong, Hong Kong

This series aims to broadly cover all the aspects related to environmental assessment of products, development of environmental and ecological indicators and eco-design of various products and processes. Below are the areas fall under the aims and scope of this series, but not limited to: Environmental Life Cycle Assessment; Social Life Cycle Assessment; Organizational and Product Carbon Footprints; Ecological, Energy and Water Footprints; Life cycle costing; Environmental and sustainable indicators; Environmental impact assessment methods and tools; Eco-design (sustainable design) aspects and tools; Biodegradation studies; Recycling; Solid waste management; Environmental and social audits; Green Purchasing and tools; Product environmental footprints; Environmental management standards and regulations; Eco-labels; Green Claims and green washing; Assessment of sustainability aspects.

More information about this series at http://www.springer.com/series/13340

Subramanian Senthilkannan Muthu
Editor

Energy Footprints of the Energy Sector

Springer

Editor
Subramanian Senthilkannan Muthu
SgT Group and API
Hong Kong, Kowloon
Hong Kong

ISSN 2345-7651 ISSN 2345-766X (electronic)
Environmental Footprints and Eco-design of Products and Processes
ISBN 978-981-13-2456-7 ISBN 978-981-13-2457-4 (eBook)
https://doi.org/10.1007/978-981-13-2457-4

Library of Congress Control Number: 2018953707

This Springer imprint is published by the registered company Springer Nature Singapore Pte Ltd.
The registered company address is: 152 Beach Road, #21-01/04 Gateway East, Singapore 189721, Singapore

This book is dedicated to:

The lotus feet of my beloved Lord Pazhaniandavar
My beloved late Father
My beloved Mother
My beloved Wife Karpagam and Daughters—Anu and Karthika
My beloved Brother
Everyone working in the energy sector to make it ENVIRONMENTALLY SUSTAINABLE

Contents

Addressing Environmental Criteria and Energy Footprint in the Selection of Feedstocks for Bioenergy Production

Iana Salim, Lucía Lijó, Maria Teresa Moreira and Gumersindo Feijoo

Abstract The search for alternatives to fossil fuel resources relies on the use of renewable bioenergy. The objective of this research is to define the most relevant sustainability criteria in the exploitation of feedstocks for bioenergy production from a life cycle perspective. Three types of biofuels were evaluated: biogas, bioethanol and biodiesel. In addition, conventional and innovative biomass sources for bioenergy will be analysed and compared. A comprehensive literature review was conducted to identify the most suitable feedstocks for each type of biofuel in terms of environmental impacts and energy-related indicators. Many studies have identified inconsistent results (from very positive to negative environmental consequences), leading to great uncertainty on this issue. Cereal crops (wheat, maize and triticale) and animal waste are examples of feedstocks for biogas production. Sugarcane, sugar beet and their by-products (molasses and bagasse) are recognised as technologically validated biomass sources for bioethanol production. As for biodiesel, oilseeds (soybean, palm, sunflower, etc.) and cooking oil residues are possible feedstocks. The number of environmental studies related to emerging biofuel feedstocks, such as algae and jatropha for biodiesel; poplar, beech, black locust for bioethanol and macroalgae for biogas, is steadily increasing. Therefore, the processes involved in feedstock production should be properly calculated for an accurate assessment, as they play an important role in the substitution of fossil fuels. The outputs from the life cycle assessment (LCA) methodology will help to support decision-making in the analysis of alternatives and avoid misleading conclusions. The Energy Return on Investment (EROI) methodology was found to be suitable

I. Salim (✉) · L. Lijó · M. T. Moreira · G. Feijoo
Department of Chemical Engineering, Universidade de Santiago de Compostela, Santiago de Compostela, Spain
e-mail: iana.camara.salim@usc.es

L. Lijó
e-mail: lucia.lijo@usc.es

M. T. Moreira
e-mail: maite.moreira@usc.es

G. Feijoo
e-mail: gumersindo.feijoo@usc.es

S. S. Muthu (ed.), *Energy Footprints of the Energy Sector*,
Environmental Footprints and Eco-design of Products and Processes,
https://doi.org/10.1007/978-981-13-2457-4_1

for comparing conventional and innovative biomass sources for bioenergy. The results vary considerably from one feedstock to another, due to differences in geographical distribution, agricultural practices and energy efficiency index. In many cases, it was found that the use of by-products (e.g. maize stover for bioethanol) or waste (pig slurry for biogas) as a biomass source for biofuels could have better EROI values than first-generation feedstocks due to the allocation of environmental burdens.

Keywords Biofuels · Energy crops · Energy Return on Investment (EROI) Life Cycle Assessment (LCA) · Second generation feedstocks · Sustainability criteria · Cumulative energy demand (CED)

1 Introduction to Bioenergy Production

It is impossible to consider life without energy. Energy has been one of the most fundamental drivers of progress and prosperity in all civilisations of human history. In the 18th century, during the Industrial Revolution, coal began to replace wood as a fuel source, marking the beginning of the fossil fuel era. Henceforth, this biomass, which has stored the energy released by the sun over geological time scales, became available to serve society for the first time in history. In the 19th century, crude oil (petroleum) was added to the energy mix, signalling the beginning of the oil era. Later, the oil crisis dramatically boosted demand for natural gas in the early 1970s, while hydroelectric, geothermal and nuclear power sources increased steadily. Over the last decade, biofuels have been extensively investigated and produced (Bithas and Kalimeris 2016).

To date, fossil fuels are the prevailing energy sources, accounting for about 80% of the global energy demand, with oil, coal and natural gas responsible for 33, 30 and 24% of the primary energy consumption, respectively (Abas et al. 2015; World Energy Council 2016). However, carbon-based fuels are non-renewable energy sources with limited reserves. It is estimated that, at existing levels of consumption, fossil fuels can last a maximum of 120 years (Guo et al. 2015).

Besides, the combustion of fossil fuels produces harmful environmental impacts, such as climate change, ozone layer depletion, acidification and air pollution (Nicoletti et al. 2015). Among them, coal is the most polluting carbon-based source, contributing to about 45% of the overall CO_2 emissions worldwide (IEA 2017a). In the current framework of growing global energy demand, actions have been taken to reduce greenhouse gas (GHG) emissions associated with the consumption of fossil fuels. In this context, Europe has set itself the target of reducing GHG emissions by 40% by 2030 compared to 1990 levels (Bonn et al. 2015). Renewable energies presented an average annual growth rate of about 5.5% between 2005 and 2015 (Eurostat 2017). Efforts to find sustainable substitutes for non-renewable energy sources have therefore intensified. In the current context of certain resources influenced by the geopolitical characteristics of the production region, diversification of energy sources will ensure a steady and secure energy supply for many countries (Sonnemann et al. 2015).

The European Commission has set targets for achieving an efficient and sustainable low-carbon economy by 2050 (Scarlat et al. 2015). It is expected that the bioeconomy will play a key role in the extension and discovery of valuable and advanced biomass in the coming decades, promoting and encouraging the efficient use of these resources. To this end, supply chain systems will need to be adapted, integrated or renewed to drive the development of a sustainable bioeconomy.

In fact, bioenergy produced from biomass has been traditionally used for millennia until today for cooking and heating (IEA 2017b). However, this practice, particularly in developing countries, has often been unsustainable due to inefficient material use, causing health and environmental issues (Lelieveld et al. 2015). Modern and sustainable sources of bioenergy are promising alternatives. In this context, favoured by technological development, the efficient conversion of biomass into solid, liquid and gaseous biofuels has been increased considerably, reducing production costs and improving its competitiveness on the market (IEA 2017b). In addition, waste and biomass valorisation can also support rural development and improve agricultural production (Scarlat et al. 2015). Figure 1 shows the development of renewable energy production in Europe. As depicted, as with other forms of renewable energies in the EU-27, biomass production is expected to increase by 2020. Bioenergy from biomass accounted for 5.9 Mtoe[1] in Europe in 2005 and is expected to increase by 19.9 Mtoe in 2020 (Beurskens and Hekkenberg 2011).

The main drivers for supporting biofuels development are: (1) they have the potential to reduce GHG emissions compared to conventional energy sources; (2) the required biomass can be grown locally; (3) countries can become less dependent on price fluctuations and feedstock imports; (4) they can be fused with oil; (5) they can support rural economic development (Lendle and Schaus 2010; Markeviius et al. 2010). However, the production of bioenergy from biomass can be a controversial issue for many reasons, including food security, land competition and use, biodiversity loss and food prices. Nevertheless, by-products of agricultural/forestry operations, industrial waste streams and abandoned, degraded and marginal lands as alternatives to fertile farmland are important options to consider when dealing with biomass for bioenergy (IEA 2017b; Immerzeel et al. 2014; Timilsina and Shrestha 2010).

Many studies in the scientific literature have evaluated the multiple possibilities of the raw materials available for bioenergy. It comprises starch crops, such as maize and wheat (Chen et al. 2018a; Muñoz et al. 2014); sugar crops, such as sugar cane and sugar beet (Duraisam et al. 2017); oil-seeds, for instance, canola (Efe et al. 2018); waste streams from industrial or domestic activities, i.e. animal waste, sewage sludge; organic fraction of solid urban waste (OFMSW) (IEA 2017b;

[1]Mtoe: Million Tonnes of Oil Equivalent. It is a standard unit with net calorific value 41 868 MJ, which is the net energy of a tonne of crude oil.

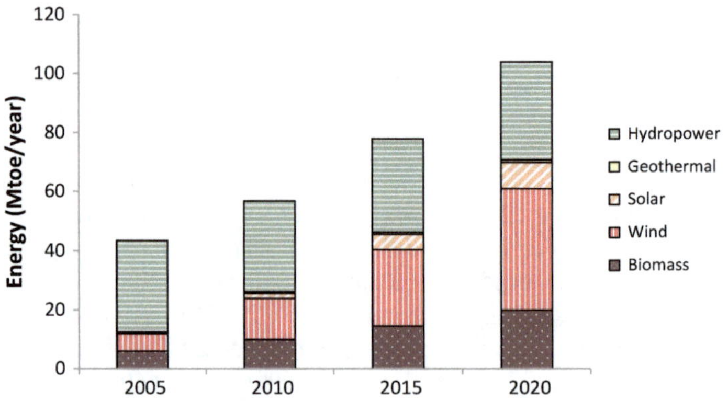

Fig. 1 Projections of energy production from renewable sources in the EU Adapted from Beurskens and Hekkenberg (2011)

Timilsina and Shrestha 2010) and kitchen oil waste (Ho et al. 2014); lignocellulosic biomass from forest operations, such as wood chips (Neupane et al. 2011) or by-products of agricultural activities or crop processing (e.g. straw, bagasse, molasses) (Nakanishi et al. 2018; Sadik and Halema 2014; Tutt et al. 2012); In addition, another perspective of feedstock source for bioenergy could be the use of macro- and micro-algae (Chia et al. 2017).

Depending on the type of feedstock and production process, biofuel technologies can be classified as first, second or third generation biofuels. First-generation or conventional biofuels, which compete directly with food, have advanced technologies and are currently being marketed. They typically use starch crops (e.g. wheat or maize) for biogas production, sugar and starch crops (e.g. maize, sugar cane, sugar beet) for ethanol production and oilseed crops (e.g. palm oil and soybean oil) for biodiesel production. Second-generation biofuels use non-edible biomass or those that do not compete directly with the food market. Certain food products may become second-generation fuels when they are no longer useful for consumption.

Many second-generation biofuels are lignocellulosic biomass, whose main composition is cellulose (Isikgor and Becer 2015). Although lignocellulosic biomass represents the most abundant carbohydrate on earth, it is currently more difficult and costly to convert cellulosic biomass into liquid fuels as an alternative to starch, sugar and oils (Timilsina and Shrestha 2010). With the exception of a few raw materials (e.g. jatropha), these processes are still in the technological and commercial development phase. The classification can be extended to third-generation biofuels, which are those produced by advanced feedstocks (e.g. algae). Unlike land-based raw materials, algae are not affected by volatile food prices and changes in land use (Chia et al. 2017). Figure 2 shows the available and potential routes of biofuel production from a life-cycle perspective.

As noted in Fig. 2, along with the variety of feedstock options, there are many types of processes that can be used and combined on the path to biofuel production.

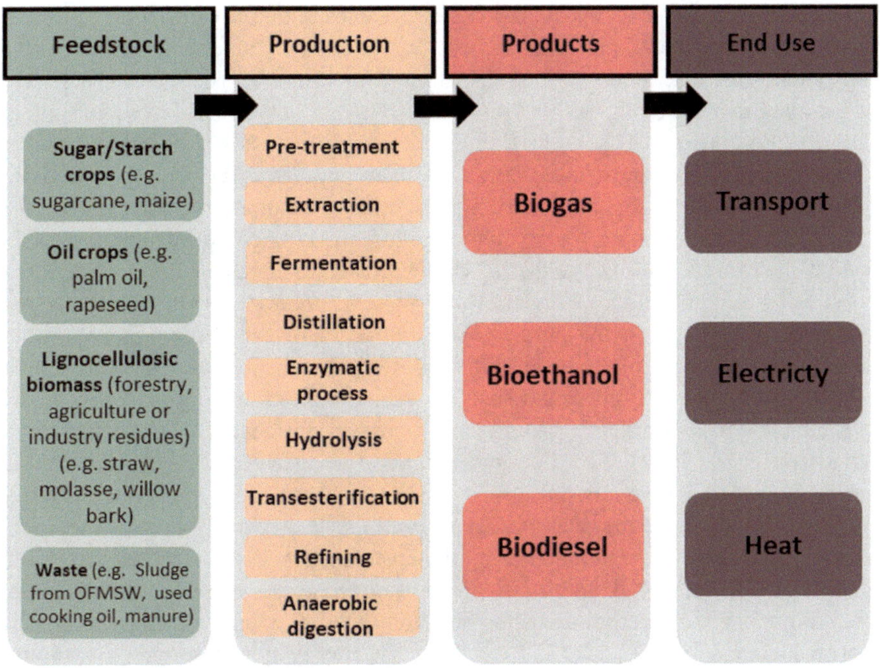

Fig. 2 Life cycle biofuels pathways

The choices already begin with how, when and where to grow and store the crops, which are important stages that will determine the quality of the raw material. The following steps, especially those relating to the initial processing stages, are decisive for efficient production and also depend on the type of biomass. The use of sugar crops requires the extraction of sucrose for subsequent fermentation (Duraisam et al. 2017). Moreover, starch crops, after the milling process (starch extraction), need to carry out a hydrolysis pathway to produce glucose prior to fermentation (Arifeen et al. 2009; Deloitte 2014). On the other hand, lignocellulosic biomass is more difficult to process, leading to a pre-treatment route prior to hydrolysis to separate its recalcitrant chemical composition into lignin, hemicellulose and cellulose (The German Federal Government 2012). Oil crops require a transesterification step that converts lipids into biofuel (Anastopoulos et al. 2009). Biogas is commonly produced in Europe in a single-stage reactor as a result of anaerobic digestion of crops such as maize and wheat, alone or in combination, for example, with waste streams such as MSW and animal manure.

Shifting from a status-quo to a new paradigm will always mean overcoming political, social, environmental and economic barriers. Sustainability assessment frameworks, including Life Cycle Assessment (LCA), use environmental, economic and social criteria and indicators to assess the performance of products and processes. Criteria add qualitative meaning and guidance to standards/certifications, while indicators are usually quantitative variables that provide information on performance, efficiency, compliance, etc. In the field of bioenergy and biofuels, many studies have defined sustainability frameworks for the selection of criteria and indicators (Buchholz et al. 2009; ISO Standard 2011; Meyer and Priess 2014; Van Dam and Junginger 2011), highlighting the relevance of GHG emissions and the energy balance. Not only are these two main indicators critical, but also the assessment of the impacts associated with land use, air emissions, water use, biodiversity loss, price and supply of raw materials, working conditions, etc. (Fritsche et al. 2012; Fritsche and Iriarte 2014; Markeviius et al. 2010).

Focusing on the energy efficiency of biofuels, there are many methods and indicators, depending on the purpose of the study. However, many indicators are not standardised and often have different designations for the same objective, making it difficult to compare studies. For example, the meaning of "energy use" does not have a standard pattern and varies considerably between reports (Arvidsson et al. 2012). The Energy Return on Investment (EROI) and the Net Energy Ratio (NER) are examples of indicators used for energy analysis related to biofuels. They evaluate the relationship between the energy distributed as fuel to the market and the energy generated during the upstream process of a product life cycle (Hall et al. 2014). EROI and NER are non-dimensional variables that facilitate communication and interpretation for decision-making. Moreover, the Cumulative Energy Demand (CED), also called "primary energy consumption", is another popular indicator that has been used for the EROI calculation (Frischknecht et al. 2015). The variables chosen for the calculation of these indicators affect the results obtained, for instance, whether or not the analysis considers `primary energy', which is the energy content inherent in the feedstock and renewable energy.

One approach to calculate the EROI is to divide the energy supplied by the non-renewable CED (Lijó et al. 2015), as described below. In this way, only the consumption of non-renewable energy is considered, as it makes it possible to quantify the amount of energy from fossil fuels actually needed for the production of biofuels in relation to the potential energy produced.

$$EROI = \frac{Energy\,delivered}{Non - renewable\,CED} \tag{1}$$

The main objective of this book chapter is to review the most relevant environmental and energy criteria in relation to the exploitation of feedstocks for bioenergy production from a life-cycle perspective. Three types of biofuels were evaluated: biogas, bioethanol and biodiesel. First, a review of the sustainability criteria was carried out to find current biomass sustainability initiatives for bioenergy. Secondly, a thorough review of the literature was conducted to identify the most appropriate feedstocks for each type of biofuel in terms of environmental impact and energy-related indicators. Finally, an energy analysis of the most commonly used feedstocks was carried out, considering the CED and EROI indicators, to compare their energy footprint.

2 General Overview on Sustainability Criteria for Feedstock

One way forward is to understand the value of sustainability, how to evaluate it and what criteria to use. In addition, studies and practices related to sustainable biomass assessment are based on non-standard guidelines, with a lack of comparability (Bosch et al. 2015). A multitude of certification standards can be found, governed by private and public organisations, focusing on the production and use of feedstocks at different scales. For instance, the International Trade Centre (ITC) unveiled a platform, in the form of a Sustainability Map, covering more than 230 voluntary standards options.[2]

Bioenergy sustainability initiatives are regulatory or voluntary schemes aimed at increasing and improving the use of biomass for energy purposes. In Europe, the most well-known initiative is the Renewable Energy Directive (RED) (RED—Directive 2009/28/EC), which targets carbon emission limits and increase the use of renewable energies in the transport sector. In the USA, the Renewable Fuel Standard (RFS-2) specifies the percentages of biofuels to be blended with their fossil counterparts. In terms of voluntary schemes, a wide range of options can be found, such as the Forest Stewardship Council (FSC) and the International Sustainability and Carbon Certification Scheme (ISCC). Table 1 below presents more examples and summarises the major biomass sustainability initiatives for bioenergy reviewed in the literature.

All these aforementioned initiatives demonstrate the growing concern for the sustainability of bioenergy and the involvement of many stakeholders in the development of guidelines for a sustainable bioeconomy. However, there is a lack of binding instructions that include, with equal importance, the three pillars of sustainability. For example, the mandatory sustainability criteria of Directive 2009/28/EC (European Parliament 2009) on the use of energy from renewable sources mainly concern environmental aspects, taking into account soil protection, biodiversity, carbon stocks, GHG savings, etc. It also includes socio-economic aspects, but is not considered relevant in this directive. However, some EU Member States, such as the *"Netherlands Technical Agreement on Sustainability criteria for biomass for energy purposes"* (NTA8080 2011) and voluntary initiatives, such as the *"Roundtable on Sustainable Biomaterials (RSB)"* (RSB 2016) provide a more detailed analysis of socio-economic considerations (Fritsche and Iriarte 2014).

The use of criteria and indicators is essential to ensure the reliability of the sustainability assessment. They work together in the same direction, with the aim of establishing conditions and requirements to ensure sustainability. Sustainability criteria guide the improvement and compliance with environmental, social and economic performance standards, which must be met by stakeholders within the biofuel production chain. However, accounting for the sustainability of biofuels is

[2]https://www.sustainabilitymap.org/home.

Table 1 Sustainability initiatives for bioenergy. Adapted from Allen et al. 2016; IEA 2017b

Activity	Brief description
Regulatory Sustainability requirements	
European Directive (RED—Directive 2009/28/EC)	One of the main targets is to reach 20% of renewable energy use by 2020 in Europe and 10% of renewable energy share for the transportation sector. Each Member State has their own national procedures to achieve this goal. It is the main directive in Europe for Biofuels and Bioenergy (European Parliament 2009). The RED established from the beginning of 2018 a threshold of at least 50% GHG emission reduction for the current and old production facilities. Nevertheless, for new facilities, which have started their activities after October 5th, 2015, the threshold is 60%. The baseline for comparison is 94.1 g CO_2 eq/MJ FBS (Fuel baseline standard) (FBS)[1]
US Renewable Fuel Standard (RFS-2)	This directive specifies the amount of biofuel, depending on its type, that must be mixed with conventional biofuels from 2006 until 2020 (EPA 2010)
The California Low Carbon Fuel Standard (LCFS)	The objective is to decrease carbon intensity from the transportation sector in California State (US) by 10% (minimum) until 2020
Oregon Clean Fuels Program (HB 2186)	The objective is to decrease carbon intensity from the transportation sector in Oregon State (US) by 10% (minimum) until 2025[3]
Canada Clean Fuel Standard	The aim is to have an annual saving of 30 Mt GHG emissions by 2030[4]
The German Biofuels Quota Act	This act establishes a minimum 6.25% of biofuels used in the transportation[5]
Pró-Álcool (Brazil)	The National Alcohol Program (Pró-Álcool) surged in 1975 due to the oil crisis period in 1973. In 1984, for instance, 84% of the cars were powered by ethanol (Kazamia and Smith 2014)
Voluntary certification schemes	
Forest certification systems	Forest Stewardship Council (FSC): It establishes sustainable practices for forest managements. Although there are some initiatives regarding wood for bioenergy and GHG balances, it is not their main priority. However, FSC includes mainly biodiversity, water, soil and human rights (FSC 2017)
Agricultural certification systems	Sustainable Agriculture Network (SAN) with Rainforest Alliance (RA): SAN/RA and GlobalGAP: both considers mainly organic certification (Rainforest Alliance,2015). There are some certifications that are restricted to a specific crop: Round Table Responsible Soy (RTRS), Roundtable on Sustainable Palm Oil (RSPO) and Bonsucro (Sugarcane). RTRS, RSPO and Bonsucro include considerations for biofuels, such as GHG emissions (Allen et al. 2016)

(continued)

Table 1 (continued)

Activity	Brief description
General biofuel/bioliquids certification systems	Most of the certifications were developed to comply with the EU RED directive. Examples of certification for biofuels and bioliquids are: International Sustainability and Carbon Certification System (ISCC)[6]; Certification for Biomaterials Biofuels and Biomass: Roundtable on Sustainable Biomaterials (RSB)(2016); Biomass Biofuel Sustainability Voluntary scheme (2BSvs) (2016) and REDcert[7], from Germany. They generally focus on downstream process and GHG emissions saving, compared to conventional fuel. However, few considerations regarding the upstream process, such as fertiliser use, tillage, soil quality, working conditions, etc. (Allen et al. 2016)
Voluntary certification schemes	
Wood pellet certification systems	Green Gold Label (GGL)[8] and the Laborelec[9]: they also can be used with other certification scheme, such as FSC. Also, the Initiative Wood Pellet Buyers (IWPB)[10] to allow the transaction of industrial wood pellets among associates
ISO standard 13065:2015	The ISO 13065:2015[11]—"*Sustainability criteria for bioenergy*" recognises the importance of waste as a potential input for carbon balance and it specifies some references of how it should be handled. It is mainly built to simplify the comparison between bioenergy sources.
NTA 8080, 2012	"Netherlands technical agreement Sustainability criteria for biomass for energy purposes" (NTA8080 2011)

[1]http://www.f3centre.se/renewable-fuels/fact-sheets/eu-sustainability-criteria-biofuels
[2]http://www.energy.ca.gov/low_carbon_fuel_standard/
[3]http://www.oregon.gov/deq/aq/programs/Pages/Clean-Fuels-History.aspx
[4]https://www.canada.ca/en/environment-climate-change/services/managing-pollution/energy-production/fuel-regulations/clean-fuel-standard.html
[5]https://www.bmwi.de/Redaktion/EN/Artikel/Energy/mineraloel-biokraftstoffe-und-alternative-kraftstoffe.html
[6]https://www.iscc-system.org/process/certification-scopes/iscc-for-energy/
[7]https://www.redcert.org/en/
[8]http://www.greengoldcertified.org/
[9]http://www.laborelec.be/ENG/services/biomass-analysis/
[10]http://www.laborelec.be/ENG/services/biomass-analysis/initiative-wood-pellet-buyers-iwpb/
[11]https://www.iso.org/standard/52528.html

not an easy task, as it depends on multiple variables, such as the type of biomass, geography, technology, consumption patterns, etc. In addition, the lack of consensus on methodologies, the numerous assumptions and the different limits of functional systems and units further challenge this assessment, leading to great uncertainty (Markeviius et al. 2010). Table 2 summarises examples of sustainability criteria for biofuels, according to the literature.

Table 2 Sustainability criteria for biofuels. Adapted from Allen et al. 2016, Lendle and Schaus 2010

Criteria	Description
General criteria	
Greenhouse gas emission intensity	To determine an absolute value to decarbonise transport fuels, instead of setting relative values to compare with fossil fuels
Land use changes and displacement effects	The biomass production used for biofuels should not incite indirect and direct dislocation of current food, feed and timber agriculture/silviculture practices in a precise zone. Special attention to avoid feedstock production on non-agriculture land of high ecosystem value
Scale of deployment	Competition with food/ feed stocks, biodiversity and ecosystem quality should not be negatively affected by the production of biofuels. It is important to set limits to the amount of land devoted for bio-based products, mainly those which use first generation feedstocks
Consistency with existing legislation	Feedstock agriculture/silviculture production, harvesting, processing, use and end-use stages should comply with international, national, regional and local mandatory rules
Specific criteria	
Resource and Energy efficiency	Resource and energy efficiency are very important issues, concerning mainly social and economic values. Energy demand and energy security must indispensably and positively walk together
Prioritisation for waste and residues	Since there are many concerns about using first generation biomass for biofuels due to land use, food competition, food price, etc, it is essential to prioritise waste, by and co-product to be used as feedstock for bioenergy
Working conditions	The worker´s quality of life is essential for assuring a fair product delivery
Biodiversity value	Feedstocks used for biofuels should not be produced on land with high biodiversity value, such as primary forest, protected areas, etc.
Carbon stock	Feedstocks used for biofuels should not be produced on land with high carbon stock, such as wetlands, continuously forested areas, etc.
Ecosystem function	Feedstocks used for biofuels should not affect ecosystem functions, such as the reduction in the diversity of species, soil erosion, decline of water quality and availability, reduction of soil fertility, etc.
Invasive species	Feedstocks used for biofuels that includes new non-native or invasive species, such as some algae, should be complemented with risk assessment study and continuous monitoring
Whole trees	Trees used for biofuel production should not be used entirely, but it should be defined according to each forest biomass, using diameter at breath height (DBH) as an indicator. DBH is a standard indicator for tree measures

3 Feedstock Selection for Bioenergy Production

In the literature, there are several LCA studies that address the environmental implications of bioenergy systems. However, the use of different methodological assumptions, such as different input data, functional units, allocation methods and reference systems, together with the use of specific local factors, hinders the comparison between environmental outcomes (Cherubini and Strømman 2011). Despite this fact, most LCA studies have found significant reductions in GHG emissions and energy consumption when biofuels are used to replace conventional energy sources (Cherubini et al. 2009). There is also evidence that biofuels cause an increase in other environmental impacts, such as acidification, eutrophication and land use. The main source of these impacts is attributed to dedicated crops, which involve the intensive use of fertilisers and pesticides, causing soil and water pollution. Therefore, the selection of biomass used as feedstock for bioenergy production is crucial for the environmental impacts produced in the life cycle of biofuels.

In terms of energy footprint, it should be considered that the cultivation of dedicated crops requires fossil energy for biomass production, mainly in agricultural or forestry operations (machinery production and diesel extraction), as well as in the production of fertilisers and other chemicals such as pesticides. Undoubtedly, the more fossil fuel required for the production of a certain amount of biofuel, the less desirable it is as a source of bioenergy (Cherubini et al. 2009). Consequently, the energy efficiency of the process is a key factor in the selection of feedstocks for bioenergy purposes, mainly due to crop yields, agricultural or forestry activities and the amount of fertilisers and other chemicals used.

3.1 Biogas

- **Feedstocks for biogas systems and environmental consequences**

Biogas is a biofuel produced, together with digestate, during the anaerobic digestion process. Anaerobic digestion of waste streams shares the principles of circular economy by converting waste into energy, water and nutrients. Biogas as a biofuel is a very versatile form of renewable energy, as it can be used for the production of heat, cogeneration of electricity and heat, as a vehicle fuel or distributed on the natural gas grid after its upgrade into biomethane (Da Costa-Gomez 2013). The digestate produced can be used as organic fertiliser because of its nutrient content, decreasing the need for mineral fertilisers and providing water, relieving pressure on limited water resources (Norton-Brandão et al. 2013). Figure 3 shows the general stages of the life cycle of a biogas system.

During the 1970s, anaerobic digestion was mainly applied to the stabilisation of waste flows (Al Seadi et al. 2013). In this regard, anaerobic digestion was considered an option for the management of large quantities of animal waste produced by the livestock sector, as well as other organic sludge streams produced by the treatment of industrial and municipal wastewater, which posed a potential risk of contamination (Holm-Nielsen et al. 2009). In contrast, the cultivation of dedicated crops, such as cereals, was developed in the 1990s in countries such as Germany and Austria (Al Seadi et al. 2013). In fact, the popularity of financial incentives for biogas production increased the use of these energy crops. For example, in Germany, about three quarters of the substrates are energy crops (mainly maize silage) (Einarsson and Persson 2017). This raised concerns about land competition, rising food prices and indirect land-use change (iLUC). As a result, some of the major biogas producers in Europe (Germany and Italy) have modified the subsidy scheme to encourage the use of agricultural and livestock waste.

According to Eurostat, the European Union produced 15.6 Mtoe of biogas in 2015: 4.2% more than in the previous year. More than 75% was concentrated in three countries: (1) Germany (7.9 Mtoe), (2) the United Kingdom (2.3 Mtoe) and (3) Italy (1.9 Mtoe). As regards the feedstock, the main biomass currently used for biogas production is energy crops, animal manure and slurry, agricultural and food wastes, as presented in Table 3. Among the different routes of energy use, biogas is mainly used for electricity production, with 60.9 TWh in 2015 (5.3% growth in just one year).

In order to ensure the cost-effectiveness of the process, the potential energy and nutrient content of the feedstock must be taken into account, which has a direct effect on the potential production of biogas, as well as on the quality of the digestate produced. In addition, there are other challenges related to legislative requirements and technological difficulties in digesting some raw materials (Feiz and Ammenberg 2017).

Energy crops are common substrates for bioenergy production due to their high biogas potential. Their cultivation requires a high input of fertilisers, pesticides and energy for agricultural and transport activities, entailing substantial environmental impacts due to emissions to air, water and soil (European Commission 2014). They can be fed immediately to the digester after harvest or stored as silage for year-round availability. Energy crops are various types of grass, cereals, beet, potato and sunflower. Among them, maize is the most widely used in Europe (European Commission 2014). The growing demand for maize may imply a change in land use, increasing the pressure to convert pastures and peatlands into areas for maize cultivation. In this regard, alternative crops such as sugar beet have recently been proposed as an alternative for bioenergy production (Jacobs et al. 2017).

However, the concern about the use of cereals for energy purposes is not only related to the environmental impacts of their production. According to Mela and Canali (2014), more than 10% of the available agricultural land in the Po Valley (Italy) was occupied by energy crops, especially maize, which could displace food crops. Since different energy crops generate different energy yields per hectare, it is essential to increase the efficiency of agricultural land use (Gissén et al. 2014).

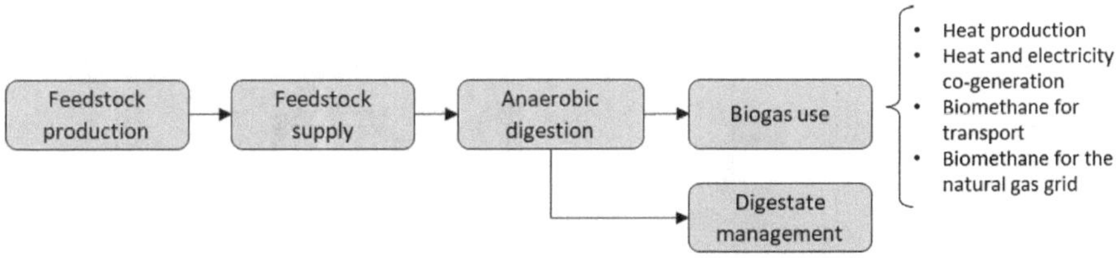

Fig. 3 Life-cycle chain of biogas systems

Table 3 List of potential and available feedstocks for biogas production

First-generation biogas	
Cereals	Barley, Rye, Triticale, Wheat
Others	Maize, sorghum
Second-generation biogas	
Animal waste	Cattle manure, pig manure, poultry manure, rabbit manure
Agricultural waste	Maize straw, triticale straw, wheat straw
Industrial and municipality waste	Food waste, green waste, sewage sludge, waste from biofuel production
Animal waste	Cattle manure, pig manure, poultry manure, rabbit manure, sheep manure
Third-generation biogas	
Microalgae	Spirulina
Macroalgae	*Sargassum spp.*

It is foreseeable that, in the future, agricultural land used for energy purposes may be limited by European regulations (Gissén et al. 2014).

In terms of waste type, there are several potential sources of biomass for biogas production (Table 4). Agricultural residues are produced as crop residues for human and animal consumption (e.g. straw). These substrates can be left on agricultural soils or used for animal husbandry (Einarsson and Persson 2017). In this sense, when left on agricultural soil, crop residues are an important source of organic matter for the soil, and their removal can lead to significant losses of biogenic carbon (Einarsson and Persson 2017).

Anaerobic digestion of manure recovers energy and reduces the risk of pathogens during land spreading (Akbulut 2012). They offer an adequate carbon/nitrogen ratio (C/N) (around 25/1) and are rich in several nutrients, necessary for the growth of endogenous anaerobic microorganisms and provide a high buffering capacity that can help stabilise the process in case of pH decrease. However, they have a low dry matter content, which negatively affects methane yield and results in high transport costs (Al Seadi et al. 2013). The yield of biogas production from manure determines the economic viability of biogas plants located in livestock production areas. Smaller and more dispersed facilities reduce emissions associated with manure transport and digester management, while better supporting local farmers' incomes (Negri et al. 2016).

Table 4 Literature review LCA studies on feedstocks for biogas production. Adapted from Morales et al. (2015)

Year	Location	Feedstock	FU	CC	AD	AC	EP	OD	PS	ET	HT	LU	FD	WD	EN	Allocation	Methods	Source
2011	Germany	Rye and maize	1 ha of crop land	x		x	x								x	N.C (system expansion)	N. D.	Bühle et al. (2011)
2012	Germany	Cattle manure, straw, maize, grass, wheat, OFMSW, paunch content, pomace, grease sludge	1 t of organic material	x	x	x	x	x	x	x	x	x	x	x		N.C (system expansion)	ReCiPe method	Poeschl et al. (2012a ,b)
2012	The Netherlands	Pig manure, maize silage, crude glycerine, beet tails, wheat yeast concentrate and roadside grass	1 t fresh matter	x		x	x					x	x			N.C (system expansion)	ReCiPe method	De Vries et al. (2012)
2012	Germany	Maize	1 kWh electricity produced	x		x	x								x	N.C (system expansion)	CML 2001 method	Dressler et al. (2012)
2013	Italy	Wheat, triticale and maize (class 300, 400, 500, 600, 700)	1 t fresh matter	x	x	x	x	x	x	x	x		x		x	N.C	CML 2001 method, and CED	González-García et al. (2013)
2013	United Kingdom	Cattle manure	1 m^3 of biogas	x		x	x			x		x	x			Mass and economic allocation between biogas and digestate	Ecoindicator 99	Mezzullo et al. (2013)

(continued)

Table 4 (continued)

Year	Location	Feedstock	FU	CC	AD	AC	EP	OD	PS	ET	HT	LU	FD	WD	EN	Allocation	Methods	Source
2014	Denmark	Pig slurry, wheat straw, organic household waste	1000 kg of pig slurry	x		x	x						x			Exergy allocation between electricity and energy	ReCiPe 1.08	Vega et al. (2014)
2014	United Kingdom	Manure, waste maize silage, fodder beet and cheese whey	1 MWh of heat and electricity co-generated	x	x	x	x	x	x	x						N.C	CML 2011 method	Whiting and Azapagic (2014)
2015	Italy	Maize, sorghum, triticale, cow slurry, sugar beet pulp, winery waste	1 MJ of net electricity produced	x	x	x	x	x	x				x			N.C	ILCD Handbook	Fantin et al. (2015)
2017	Italy	Maize, triticale, pig slurry, chicken manure, organic fraction of municipal solid waste, food waste	1 MWh of electricity produced	x	x	x	x	x	x				x			N.C (system expansion)	ReCiPe	Lijó et al. (2017)
2017	United Arab Emirates	Marine algae and cattle manure	1 GJ of biogas produced	x	x	x	x	x		x		x	x			No	Ecoindicator 99	Giwa (2017)

Acronyms FU Functional Unit; *CC* Climate Change; *EN* Energy; *AD* Abiotic Depletion; *AC* Acidification; *EP* Eutrophication; *OD* Ozone Depletion; *PS* Photochemical Smog; *ET* Ecotoxicity; *HT* Human Toxicity; *LU* Land Use; *FD* Fossil Depletion; *WD* Water Depletion; *N.C.* Not considered; *N.D.* Not determined. For simplification, the impact categories in the table above are grouped: Global Warming Potential (*GWP*) comprises Climate Change (*CC*); Ecotoxicity (*ET*) and Eutrophication (*ET*) include all their derivations (terrestrial, marine, freshwater etc). Land Use (*LU*) includes also Agriculture Land Occupation (*ALO*), urban land occupation (*ULO*) and Natural Land Transformation (*NLT*). Moreover, the category "energy" (*EN*) includes net energy balance (*NEB*); net energy ratio (*NER*), cumulative energy demand (*CED*) and non-renewable energy use (*NREU*)

The use of OFMSW from separate collection provides a clean, high-quality material for anaerobic digestion that does not compete for land use and reduces organic material going to landfills or incineration. The rationale behind the use of OFMSW as an organic substrate for anaerobic digestion is its high methane potential. It should be noted, however, that the composition of OFMSW varies according to regions and seasons of the year, as well as to different collection systems. It is essential to ensure a high degree of purity, since the presence of inappropriate foreign materials, such as plastics and metals, can adversely affect process performance and digestate quality as a fertiliser. The impurity content depends to a large extent on the human factor, i.e. the awareness and motivation of the population involved in the collection systems (Al Seadi et al. 2013).

There are different studies that analyse the environmental impacts of energy crops. Bühle et al. (2011) compared the production of biogas and pellets from residual agricultural biomass with the digestion of whole crops. As shown in Table 4, the crops studied were rye and maize, growing in a double cropping system. The limits of the system studied included the release of carbon dioxide from the soil and heat production. The study included the assessment of technical indicators such as process efficiency and primary energy requirements as well as environmental indicators such as climate change, acidification and eutrophication. For energy efficiency, the authors considered the final energy supplied by biogas and pellets as produced energy and the non-renewable primary energy required in the process as an energy input. The results showed that co-production of biogas and pellets from residual biomass was a more energy-efficient option than the digestion of whole crops. In addition, GHG savings were achieved in all scenarios because of the avoidance of fossil fuel-based processes related to heat production, but no significant differences were found between the two scenarios. Finally, emissions that produce impacts related to acidification and eutrophication were higher in both cases than in the fossil reference system. The main drivers of these impacts were emissions from the combustion of biogas and pellets, as well as from the application of fertilisers.

Another paper analysed the effect of local and climatic conditions and agricultural procedures on the cultivation of maize for biogas production in Germany (Dressler et al. 2012). The system boundaries also included credits for the production of digestate and heat. The impacts studied were climate change, fossil energy demand, acidification and eutrophication. The results showed the important effect of regional factors on the environmental impacts of maize cultivation. The main reasons for these differences were different demand for fertilisers, pesticides and field work due to different irrigation needs.

Subsequent work analysed the production of biogas from maize, triticale and sorghum, as well as agricultural residues and livestock effluents (Fantin et al. 2015). The study demonstrated the importance of energy crop cultivation in the environmental performance of biogas systems. These authors also pointed out the relevant role of digestate management, which entails liquid and solid storage and their application as well as the methodological difficulties in accounting for emissions from the application of mineral and organic fertilisers in LCA studies.

Therefore, according to the studies analysed, the identification of the best energy crop for biogas production depends on different factors, including biomass yield, the requirements of fertilisers and pesticides, as well as the agricultural activities involved. In addition, double-crop systems (maize and wheat rotation or triticale) achieved better environmental results than single-crop systems.

Beyond energy crops, different studies were carried out on the production of biogas from waste. The studies conducted by Mezzullo et al. (2013), De Vries et al. (2012) and Vega et al. (2014) focused on the evaluation of animal waste for biogas production. The first studied the environmental impacts of biogas production from cattle waste at a small-scale plant. The results showed that it led to a reduction in GHG emissions and energy impact compared to kerosene-based energy. However, other environmental impacts, such as ammonia emissions, were identified. The second study focuses on the comparison of pig manure mono-digestion with the co-digestion with several possible co-substrates, including maize silage (alone or in combination with glycerine), beet tails, wheat yeast concentrate and roadside grass. The single digestion of animal waste was identified as a limited source of bioenergy; while co-digestion with other substrates increased the environmental impacts such as climate change, marine eutrophication and marine use due to the use of products for the substitution of co-substrates. Similarly, Vega et al. (2014) also evaluated possible co-substrates for animal slurry. The substrates chosen were straw, organic household waste and the solid fraction of separated slurry, and the reference scenario selected was the management of animal slurry without biogas production. Straw was identified as the most successful co-substrate due to its high methane potential and low nutrient content.

In the field of co-digestion of energy crops in combination with food waste, Whiting and Azapagic (2014) analysed the production of biogas from agricultural waste streams such as manure, cheese whey, silage maize residues and fodder beet. The study concluded that the use of agricultural residues for biogas production reduced GHG emissions; however, the use of energy crops such as maize reduced other impacts due to higher biogas yields. In a recent publication, Lijó et al. (2017) assessed the environmental consequences of replacing energy crops with food waste. The authors also analysed the influence of raw material composition on the yield of the entire biogas plant. In biogas systems using energy crops, one of the most important hot spots was the cultivation step due to diesel consumption and emissions from fertiliser application. However, these substrates had a higher biogas potential. Moreover, the characterisation of food waste demonstrated that it can be used as an alternative co-substrate capable of improving the environmental profile of biogas production due to its higher energy potential than other waste (e.g. pig slurry).

Giwa (2017) studied the suitability of marine algae (i.e. third generation feedstock) as a substrate for biogas production compared to cattle manure. According to the results obtained, macro-algae achieved slightly better results due to lower energy and water requirements than cattle manure.

- **Energy of different biogas systems**

As indicated in the previous section, the use of waste for biogas production is beneficial from an environmental point of view, since, in addition to the benefits of its valorisation for the production of bioenergy and biofertilisers, the advantages of waste management are added. In fact, in many LCA studies, the environmental burden of waste production is not considered within the boundaries of the biogas system, as its production is not affected by its valorisation as bioenergy. Therefore, in order to calculate the energy life cycle balance of these systems, the input of energy from waste streams is limited to the collection of waste, its transport to the biogas plant and the necessary pre-treatment operations. As a result, the energy balance is very positive, even when some waste streams, such as animal waste, have a lower energy potential compared, for example, to maize silage. The situation is completely different when calculating the energy balance of energy crops. Since this biomass is grown exclusively for bioenergy purposes, all the energy needed for its cultivation, including pesticides, fertilisers, diesel fuel and the production of machinery, should be accounted for.

Aiming at analysing the energy efficiency of different substrates for biogas production, three studies on energy crops were selected (Bühle et al. 2012; Gissén et al. 2014; González-García et al. 2013). All the non-renewable energies needed for agricultural and chemical activities were accounted for to meet the fossil energy needs of different energy crops in different European countries (Germany, Italy and Sweden). In addition, pig slurry and MSW collection were considered as waste streams included in the analysis (Lijó et al. 2017). The potential biogas yield reported in the literature was used as a reference to estimate the energy production of these raw materials.

Different biomass yields and biogas potential are documented among the different feedstocks in different European countries (Table 5). Sugar beet was identified as the most productive feedstock, followed by maize cultivated in Germany and Italy. The biogas potentials of other Italian crops (wheat and triticale) together with rye produced in Germany, were also high, being hemp as the species with the lowest yield reported. It is worth noting the significant difference in potential biogas production between maize produced in Germany and Italy, according to the scientific literature, which may be caused by different types of maize, climatic conditions or management practices. Agricultural activities can be classified into field preparation, crop growth and biomass harvesting. Within field preparation, the most common activities are ploughing, harrowing and organic fertilisation with digestate. Irrigation is only carried out during maize cultivation in Italy, probably due to the drier climate compared to Germany. Mineral fertilisation is also performed on all crops under study, but with different chemicals and fertilisation rates.

Figure 4 shows the energy requirements and the production of different feedstock in different European countries, together with the EROI obtained. As shown, all these differences discussed before are translated into different energy inputs and outputs for different feedstocks. Energy crops that require more energy in their cultivation (i.e. rye and maize) also translate into higher energy production due to

Table 5 Feedstocks considered in the energy analysis

Feedstock	Source	Location	Feedstock yield	Biogas potential
Rye	Bühle et al. (2011)	Germany	32.2 t rye/ ha	108.0 L_N CH_4/kg rye
Maize	Bühle et al. (2011)	Germany	60.0 t maize/ ha	71.4 L_N CH_4/kg maize
Wheat	González-García et al. (2013); Negri et al. (2014)	Italy	37.0 t DM/ ha	78.45 L_N CH_4/kg wheat
Triticale	González-García et al. (2013); Negri et al. (2014)	Italy	38.0 t DM/ ha	92.9 L_N CH_4/kg triticale
Maize	González-García et al. (2013); Negri et al. (2014)	Italy	50.0 t DM/ ha	125.6 L_N CH_4/kg maize
Hemp	Gissén et al. (2014)	Sweden	24.8 t hemp/ha	74.7 L_N CH_4/kg hemp
Sugar beet	Gissén et al. (2014)	Sweden	97.7 t sugar beet/ha	76.9 L_N CH_4/kg sugar beet
Pig slurry	Lijó et al. (2017)	Italy	–	13 L_N CH_4/kg pig slurry
OFMSW	Lijó et al. (2017)	Italy	–	66 L_N CH_4/kg OFMSW

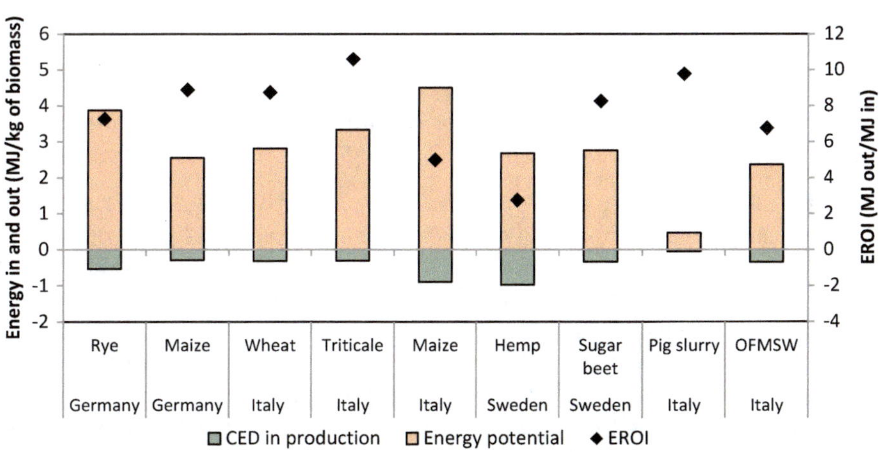

Fig. 4 Energy required and produced for each feedstock under study

the high potential for biogas production. Hemp and sugar beet have similar energy potential; however, the energy required to produce 1 kg of sugar beet is less than that required to produce 1 kg of hemp, even though sugar beet production requires more intensive agricultural practices. More specifically, while hemp requires harrowing twice a year, sugar beet includes harrowing repeated 3 times; in addition, unlike hemp, sugar beet also includes the application of pesticides (twice) and inter-row weeding (twice). Higher sugar beet yields per hectare significantly improved the energy efficiency of the crop.

As regards waste streams, pig slurry for biogas generation requires a very low energy input, since only the tractor used for its collection and transport to the plant is included; however, this feedstock produces less biogas than any other biomass under consideration (Table 3). On the other hand, OFMSW represented a higher energy input compared to pig slurry due to the impact associated with municipal waste collection; however, the biogas potential of this feedstock is much greater compared to pig slurry.

However, it should be noted that food waste streams vary greatly from season to season and from region to region. The EROI value of each feedstock, as well as the energy consumed in production and the energy potential can be found in Table 6. According to the results obtained, the highest EROI corresponds to triticale, followed by pig slurry. While triticale yields a relatively high energy yield with low energy requirements, the satisfactory results of pig slurry are motivated by the low energy demand required for its collection. Finally, hemp led to the worst ratio between the energy needed for production and the energy produced per unit mass. These results are related to a low biomass yield per hectare of cultivated land, despite relatively intensive agricultural activities.

3.2 Bioethanol

- **Available feedstocks for bioethanol production and their environmental implications**

Ethanol is a volatile, transparent and flammable liquid that is widely used as a solvent, as an ingredient in alcoholic beverage companies and as a fuel additive in the transport sector. Within the total liquid fuels available on the world market, demand for bioethanol has grown considerably and currently represents the main commercially available biofuel. World ethanol production has increased from 28.5 billion litres in 2004 to 98 billion litres in 2015 (Debnath et al. 2017). Some countries, such as Brazil and the United States, have established minimum volumes of biofuel production as government policies (Timilsina and Shrestha 2010). These two countries are leaders in the global bioethanol market, accounting for 85% of total world production of 101 billion L in 2016 (IEA 2017b).

Sugarcane produced in Brazil and maize in United States are the main sugar and starch crops used for bioethanol production, respectively. In addition to sugar cane, other sugar crops for ethanol production include sugar beet and sweet sorghum (Machado et al. 2017). For sugar crops, the process usually involves sugar extraction, fermentation and distillation (Fig. 5). Although it can be produced by chemical processes, fermentation is the most common used step, which occurs when yeasts break down sugars into alcohol and carbon dioxide (Timilsina and Shrestha 2010).

When starches are used for ethanol production, after the milling stage to extract the starch, an enzymatic process is required to convert starch into simple sugars (i.e. glucose) (Fig. 6). Therefore, this process implies increased energy use and higher costs (Timilsina and Shrestha 2010). The most commonly used starch crops for bioethanol are maize and wheat. Others include cassava, rice, barley, wheat, etc. (Machado et al. 2017).

The conversion of edible biomass for bioethanol production is increasing the debate on the real sustainability of using virgin sugar and starch crops (Bansal et al. 2016). Unlike first-generation ethanol (sugars and starches), lignocellulosic ethanol has the potential to use a wide variety of biomass sources, which do not compete with food and feed markets, found in agricultural by-products, forestry operations as well as residues from industrial and household streams.

Lignocellulosic biomass is composed of tree natural polymers: cellulose, hemicellulose and lignin. The process of conversion to sugars usually includes the following processes: pre-treatment, hydrolysis (to break down cellulose into glucose), fermentation (to convert sugars into ethanol) and purification (i.e., distillation) (Chia et al. 2017; Song et al. 2018; Timilsina and Shrestha 2010) (Fig. 7). However, the techno-economic feasibility for the use of lignocellulosic ethanol is

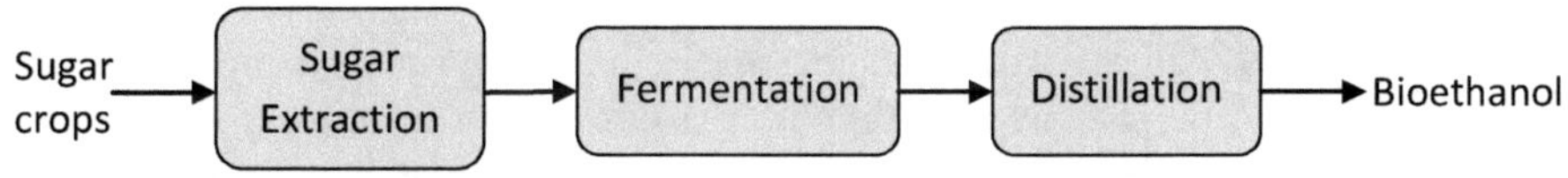

Fig. 5 Bioethanol production process from sugar crops

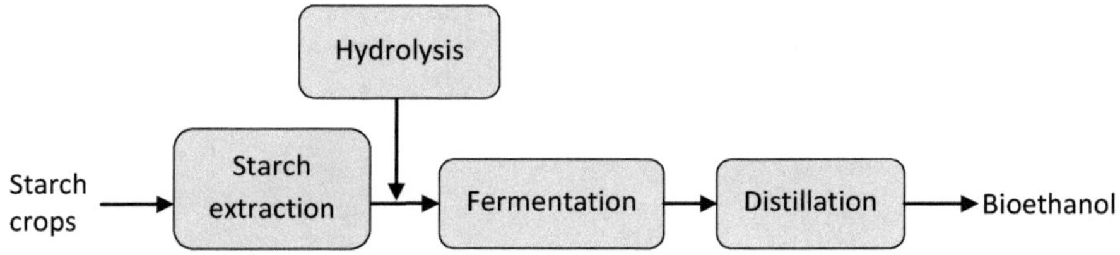

Fig. 6 Bioethanol production process from starch crops

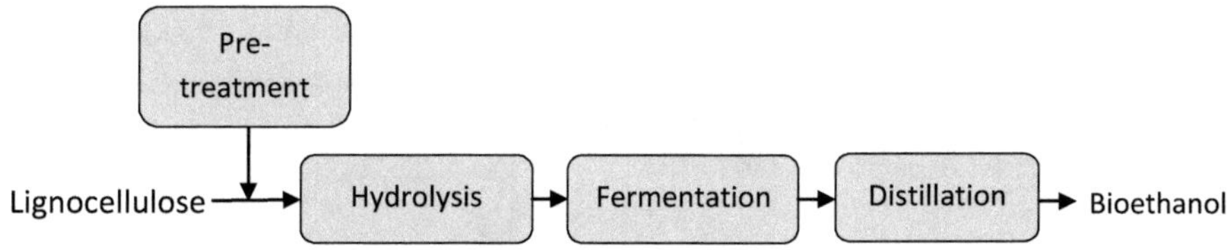

Fig. 7 Bioethanol production process from lignocellulose

now significantly lower than the process associated with the use of sugars and starch-based crops (Song et al. 2018). Today, the lignocellulosic raw materials with the greatest potential for bioethanol production are maize stover, straw, grass and bagasse (IEA 2017b; Lynd et al. 2017).

Although not yet technically and economically viable, bioethanol production from microalgae (e.g. spirulina) (Hossain et al. 2015) and macroalgae (e.g. Sargassum) (Borines et al. 2013) is very promising, as these feedstocks can store a large amount of carbohydrates, consisting mainly of starch and cellulose (Sivaramakrishnan and Incharoensakdi 2018). Table 6 below summarises the potential and available raw materials used for bioethanol.

Table 7 presents a review of the literature on LCA of different types of feedstocks for bioethanol production. It can be observed that the most common functional units (FUs) are mass of ethanol (in kg), energy content in ethanol (in MJ) or distance (in km). The most widely used impact categories are climate change, acidification, eutrophication, ozone depletion, photochemical oxidant formation, human toxicity and ecotoxicity. LCA studies on bioethanol feedstocks usually analyse both energy content (MJ_{EtOH}) and GHG emissions (kg CO_2 eq) to understand the relationship between GHG emissions and energy use (kg CO_2 eq/ MJ_{EtOH}).

The literature review summarised in Table 7 shows a time scale of approximately 10 years, which includes LCA studies of bioethanol production using first and second generation feedstocks. The results of LCA can vary considerably, depending on many factors: data quality, the impact assessment methods chosen, geographical location, climatic conditions, consideration of land-use change and biogenic carbon dioxide, system boundaries, type of feedstock, agriculture/forestry management, and so forth.

The work of Zimbardi and Cerone (2007), comparing the most common feedstocks used for ethanol production (maize, sugar cane, beet and wheat), concluded that sugar cane has by far the best energy and environmental performance, followed by maize. Sugar beet and wheat, however, had the lowest energy yield and the

Table 6 List of potential and available feedstocks for bioethanol production

First generation bioethanol	
Sugar crops	Sugarcane, sugar beet, sweet sorghum
Starch crops	Maize, rice, barley, wheat, cassava
Second generation bioethanol	
Lignocellulose from agriculture and silviculture production and residues	Corn stover, straw, bagasse, Molasse switchgrass, Ethiopian mustard, Flax shives, Poplar, miscanthus, Fiber Sorghum, Banana rachis
Industrial and commercial waste	Food waste; Pulp and paper sludge
Animal waste	Cattle manure
Third generation bioethanol	
Microalgae	Spirulina
Macroalgae	*Sargassum spp.*

Table 7 Literature review LCA studies on feedstocks for bioethanol production. Adapted from Morales et al. (2015)

Year	Location	Feedstock	FU	CC	AD	AC	EP	OD	PS	ET	HT	LU	FD	WD	EN	Allocation	Methods	Source
2007	Global	Sugarcane, maize, sugar beet, wheat	MJ	x	x	x	x	x								N. C.	CML	Zimbardi and Cerone (2007)
2010	Spain	Alfalfa stems, poplar, Ethiopian mustard, flax shives and hemp hurds	Km	x		x	x		x				x			N. C.	CML	González-García et al. (2010)
2010	Global	Switchgrass	km	x	x	x	x	x	x	x	x					Energy and economic	CML	Bai et al. (2010)
2011	Greece	Sugar beet	35Gcal	x		x	x	x		x		x				N. D.	Eco-indicator 99 and CML	Foteinis et al. (2011)
2012	US, Brazil	Maize, sugarcane, corn stover, switchgrass and miscanthus	MJ	x											x	Energy and displacement	GREET	Wang et al. (2012)
2012	Europe	Wheat straw	km	x		x	x	x	x	x			x	x		Mass	ReCiPe	Borrion et al. (2012)
2013	UK	Wheat straw	km	x	x	x	x	x	x	x	x					Economic	CML	Wang et al. (2013)
2014	US, France, Brazil	Maize, maize stover, sugarcane, sugar beet and wheat	kg	x		x	x		x			x				Economic	ReCiPe	Muñoz et al. (2014)
2014	US	Food waste	L	x												N.C.	GREET	Ebner et al. (2014)
2014	India, Brazil	Sugarcane	kg	x			x							x	x	Economic	Impact 2002+	Tsiropoulos et al. (2014)
2015	India	Molasse	t	x											x	Mass, energy, economic	Own calculation and literature and NER.	Soam et al. (2015)

(continued)

Table 7 (continued)

Year	Location	Feedstock	FU	CC	AD	AC	EP	OD	PS	ET	HT	LU	FD	WD	EN	Allocation	Methods	Source
2016	Portugal	Pulp and paper sludge	MJ	x	x	x	x	x	x	x	x					N. C.	CML	Sebastião et al. (2016)
2017	Malaysia	Cassava	L	x											x	N. C.	Own calculation and literature. NEB and NER	Hanif et al. (2017)
2017	Brazil	Cattle Manure	1000 kg	x		x	x	x	x	x	x	x	x	x		N. C.	ReCiPe	de Azevedo et al. (2017)
2017	Italy	Fiber sorghum	km	x		x	x	x	x				x	x		N. D.	ReCiPe	Forte et al. (2017)
2018	Ecuador	Banana rachis	MJ	x		x	x		x				x		x	N.C.	ReCiPe, CED, NEV and NER.	Guerrero and Muñoz (2018)
2018	US	Sugar beet	MJ	x											x	N.C.	GREET and own calculations	Alexiades et al. (2018)

Acronyms FU Functional Unit; *CC* Climate Change; *EN* Energy; *AD* Abiotic Depletion; *AC* Acidification; *EP* Eutrophication; *OD* Ozone Depletion; *PS* Photochemical Smog; *ET* Ecotoxicity; *HT* Human Toxicity; *LU* Land Use; *FD* Fossil Depletion; *WD* Water Depletion; *N.C.* Not considered; *N.D.* Not determined. For simplification, the impact categories in the table above are grouped: Global Warming Potential (*GWP*) comprises Climate Change (*CC*); Ecotoxicity (*ET*) and Eutrophication (*ET*) include all their derivations (terrestrial, marine, freshwater etc). Land Use (*LU*) includes also Agriculture Land Occupation (*ALO*), urban land occupation (*ULO*) and Natural Land Transformation (*NLT*). Moreover, the category "energy" (*EN*) includes net energy balance (*NEB*); net energy ratio (*NER*), cumulative energy demand (*CED*) and non-renewable energy use (*NREU*)

worst environmental burdens. However, local considerations are likely to influence the environmental outcomes. In India, sugar cane-based ethanol appears to be more environmentally friendly than Brazil in terms of GHG emissions, since ethanol in India is produced exclusively from a co-product of sugar cane: molasses. However, India has lower sugar cane yields and much higher irrigation water requirements (Tsiropoulos et al. 2014).

Soam et al. (2015) considered different allocation methods to assess the performance of molasses-based ethanol in different regions of India. Sugar cane cultivation alone is responsible for more than 60% of total energy consumption due to high fertiliser and water use. The results showed that the inclusion of the allocation is imperative, due to the value of the by-products of the sugar cane industry. Muñoz et al. (2014) carried out an evaluation among different feedstocks (sugarcane, maize, maize stover, sugar beet and wheat) in France, Brazil and the USA and concluded that, from a GHG perspective, biomass-based ethanol is advantageous compared to carbon-based counterpart. Nevertheless, when other environmental indicators, such as land use, are investigated, there are many trade-offs. The authors also require harmonising calculation methods related to land use change.

Wang et al. (2012) investigated maize, corn stover, switchgrass and miscanthus (cultivated in the USA) and sugarcane (cultivated in Brazil) for bioethanol production. For all potential biomass assessed, it was concluded that GHG emissions are highly dependent on the use of nitrogen fertilisers, due to the release of nitrous oxide into the atmosphere. Another important consideration is that the seasonal characteristic of the lignocellulosic cultivation. Cassava fuel ethanol showed a 73% reduction in gasoline-related GHGs (Hanif et al. 2017). From an energy point of view, the outcomes from NEB and NER showed positive values from cassava-based ethanol, confirming its energy efficiency.

With regard to sugar beet ethanol, its carbon intensity in California was estimated to be 71% lower than that of gasoline (Alexiades et al. 2018). However, regarding other impact categories, such as land-use change, conventional gasoline is preferable. Foteinis et al. (2011) assessed the impact of converting old sugar beet plants into novel bioethanol plants in Greece. The results showed that this transformation would reduce overall environmental impacts by approximately 30%. In addition, the existing infrastructure and knowledge of the sugar beet plant facilitates the redesign of the sugar beet bioethanol plants.

Bioethanol produced from switchgrass (Bai et al. 2010) and wheat straw (Borrion et al. 2012) achieved 65% and 73% reductions in GHG emissions driven by a blend of ethanol and E85 fuel, respectively. For E15 wheat straw ethanol, GHG emissions were reduced by only 13%. The E100 high-level ethanol blend was evaluated in the work of Wang et al. (2013) using five types of pretreatment technologies for wheat straw. Compared to their gasoline counterparts, the results showed a GHG emission savings of 45% using E100 ethanol. In addition, the authors indicated that enzyme production is a major burden on the environment and that, although wheat straw is a by-product, it contributes significantly to global warming, eutrophication and ecotoxicity.

In LCA studies, it is important to consider that ethanol fuel is commonly blended in the market with gasoline, for example, as a low level blend of E10 ethanol (10% anhydrous ethanol and 90% gasoline) or a high level blend of E85 ethanol (85% anhydrous ethanol and 15% gasoline). When comparing different potential lignocellulosic feedstocks (alfalfa stems, poplar, Ethiopian mustard, flax shives and hemp hurds) for ethanol production, Ethiopian mustard showed a 145% reduction in GHG emissions compared to conventional gasoline regarding the blend E85 (González-García et al. 2010). On the other hand, the amount of fuel required to drive 1 km is estimated as 0.063 kg for E10 and 0.092 kg for E 85 (Forte et al. 2018; González-García et al. 2010; Knoll et al. 2009).

The LCA study on banana lignocellulosic waste for ethanol in Ecuador showed an emission of 0.031 kg CO_2/MJ of bioethanol, compared to 0.088 kg CO_2/MJ for conventional gasoline (Guerrero and Muñoz 2018). Ecuador is the largest exporter of bananas in the world, representing a great opportunity to use the products of its crops for the production of ethanol, such as Banana rachis. In Italy, lignocellulosic fibre sorghum fuel ethanol demonstrated environmental improvements for climate change, ozone and fossil fuel depletion, but worse values for acidification and eutrophication (Forte et al. 2017). This LCA included the use of marginal land for sorghum production with the aim of reducing risks related to land competition, saving energy and reducing environmental pressure.

LCA studies on side streams from industrial or households activities for bioethanol production were also evaluated. Bioethanol conversion from food waste was assessed, with a reported improvement in GHG emissions of about 500% compared to maize ethanol and conventional gasoline, when avoided emissions from food waste disposed in landfills were included in the analysis (Ebner et al. 2014). In terms of cattle manure for bioethanol, the recovery of waste offsets the environmental burden of biofuel production. Overall, all the impact categories showed low values, being the drying process of manure the largest contributor (de Azevedo et al. 2017). Sebastião et al. (2016) modelled a pulp and paper sludge bioethanol plant. The results demonstrated that hydrolysis and neutralisation of $CaCO_3$ account for up to 85% of the global environmental impacts.

- **Energy requirements of these feedstocks**

As indicated in the previous section, the use of waste and by-products from agricultural activities is advantageous from an environmental point of view, provided that it does not affect soil quality and fertility. The number of studies on second-generation biomass, such as lignocellulosic raw materials from agricultural and forestry residues, has increased considerably over the last decade and has shown that it is possible to reduce environmental burdens.

Aiming to assess the life-cycle energy of different biomass located worldwide, this study will analyse the energy efficiency by comparing the following feedstocks: sugarcane, maize, wheat straw, maize stover and switchgrass (Bai et al. 2010; Borrion et al. 2012; Giuntoli et al. 2013; Luo et al. 2009; Muñoz et al. 2014; Tumuluru 2015), as shown in Table 8.

Table 8 Biomass considered for energy analyses for bioethanol

Feedstock	Source	Location	Feedstock yield	Bioethanol potential
Sugarcane	Muñoz et al. (2014)	Brazil	82,697 kg/ha	7.7 kg/kg EtOH
Maize stover	Luo et al. (2009); Muñoz et al. (2014); Tumuluru (2015)	United States	4,676 kg/ha	3.9 kg/kg EtOH
Wheat straw	Borrion et al. (2012); Giuntoli et al. (2013); Muñoz et al. (2014)	France	2972 kg/ha	5 kg/kg EtOH
Switchgrass	Bai et al. (2010); Mitchell et al. (2012); Tumuluru (2015)	Global	16 000 kg/ha	3.8 kg/kg EtOH
Maize	Muñoz et al. (2014)	United States	9703 kg/ha	2.5 kg/kg EtOH

As shown in Table 8, sugarcane biomass has the most productive yield, followed by switchgrass, maize, maize stover and wheat straw. Nevertheless, in terms of bioethanol potential, sugarcane has the lowest percentage, followed by wheat straw, maize stover, switchgrass and maize. Chemical fertilisation and phytosanitary application are taken into account in all feedstocks. The irrigation phase is only included for maize, maize stover and wheat straw. Only agricultural activities, harvesting and transport are investigated in this study. Additionally, an average distance of 30 km was estimated between the agricultural field and the bioethanol plant.

Figure 8 shows the calculation of EROI for the different biomass and bioethanol potentials. The energy considered for ethanol is based on its low heating value (LHV), which is 28 MJ/kg of ethanol (Rocha et al. 2014). As regards the CED calculation, only non-renewable energies used in the agricultural phase were considered.

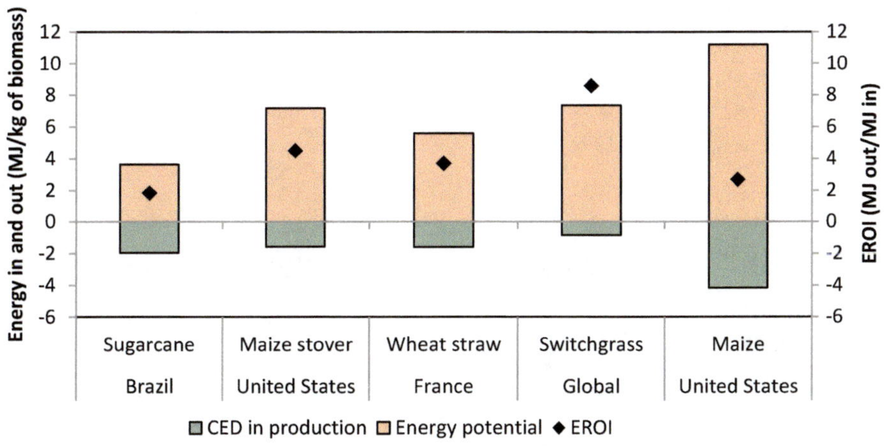

Fig. 8 Energy required and produced for each feedstock under study

As noted, maize biomass has the greatest energy potential. However, it turned out to have the highest CED. Although sugarcane is known for its high production yields, the production of sugarcane bioethanol requires more biomass inputs than maize. The maize stover is a by-product and, as a result, the fossil energy used will be reduced as the energy flows between maize and the stover are allocated. The same reasoning applies to wheat straw, as a by-product of wheat. Switchgrass showed the lowest CED because of its agricultural practices, which do not require a large amount of chemical inputs compared to traditional crops.

The biomass switchgrass achieved the highest EROI, followed by the maize stover, wheat straw, maize and sugarcane, as shown in Fig. 8. It can be observed that second-generation bioethanol has achieved a higher EROI result than first-generation bioethanol (i.e. sugar cane and maize). Compared to sugarcane, maize has a high CED value due to high irrigation practice. However, the bioethanol yield of maize is much higher than that of sugarcane, resulting in a more positive EROI. Overall, all feedstocks reached an EROI higher than 1, which means that the production of bioethanol from these raw materials is not an energy sink. However, to be considered sustainable, the literature indicates that the EROI value must have a minimum value of 3 (Carneiro et al. 2017).

3.3 *Biodiesel*

- **Environmental consequences of different feedstocks for biodiesel generation**

Biodiesel fuel is constituted by lipids, known as fatty acid methyl esters (FAME). Its production is produced mainly by transesterification, where a catalysed chemical reaction occurs between an alcohol and a vegetable oil, generating FAME and glycerol (Fig. 9) (Anastopoulos et al. 2009). Biodiesel reached a worldwide production of 36 billion litres in 2016 (IEA 2017b). Unlike bioethanol, which has a

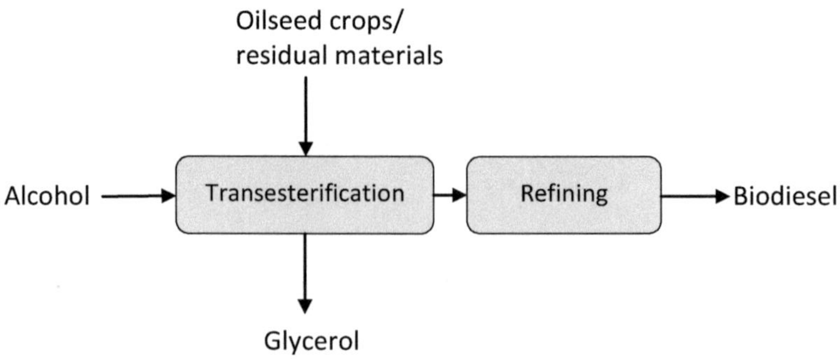

Fig. 9 Biodiesel production process

market dominated by two countries (Brazil and the USA), the distribution of biodiesel is more dispersed throughout the world. There is a wide range of possible feedstocks for biodiesel production, with more than 350 types of biomass (Oh et al. 2012; Rehan et al. 2018). This fuel is mainly produced from oilseed crops (e.g. palm oil, rapeseed, soybean, sunflower, jatropha and coconut). With the exception of jatropha, these raw materials mentioned above are classified as first-generation biodiesel, since they compete with food and feed markets (Timilsina and Shrestha 2010).

The opportunity to use second-generation biofuels has grown on the basis of the many examples found in the literature. The feedstock possibilities are those produced by non-food oilseed crops such as jatropha (Siregar et al. 2015) and camelina (Tabatabaie et al. 2016), biomass from agricultural and forestry residues, such as rice straw (Zheng et al. 2012), tomato seeds (Giuffrè et al. 2016), industrial and household waste streams, such as sewage sludge (Chen et al. 2018b), used cooking oil, animal fat waste, etc (Seber et al. 2014). As for third-generation biodiesel, efforts to produce biodiesel from micro and macroalgae have increased in the last decade (Gnansounou and Kenthorai Raman 2016; Khan et al. 2017; Kligerman and Bouwer 2015; Piloto-Rodríguez et al. 2017). Table 9 depicts examples of feedstocks for biodiesel production.

It is estimated that about 75% of the global production cost of biodiesel is due to the selection of the feedstock. In addition, it is cheaper to produce diesel from fossil fuels, between 1.5 and 3 times (Rehan et al. 2018). It is therefore important to find alternatives to the traditional oilseed crops used for biodiesel technology, as well as to improve the processing technologies of feedstocks that can compete with fossil diesel. A literature review based on LCA studies of feedstocks for biodiesel production was conducted using references from the last decade (Table 10). Two main functional units were used: mass-based (e.g. kg) and energy-based (e.g. MJ). Unlike bioethanol, few studies use km as the functional unit. This may be due to the biodiesel market as transportation fuel, which is less established and distributed than bioethanol. It is important to note that results may vary significantly according

Table 9 List of potential and available feedstocks for biodiesel productions

First-generation biodiesel	
Edible oil crops	Palm, rapeseed, sunflower, soybean, Linseed, Cotton seed
Second-generation biodiesel	
Non-edible oil crops/ Lignocellulosic residues	Jatropha, camelina, pongamia, neem, rice straw, castor oil
Industrial and commercial waste	Food waste, used cooked oil, sewage waste
Animal waste	Fish oil, tallow, poultry fat
Third-generation biodiesel	
Microalgae	*Chlorella vulgaris*, *Nannochloropsis* sp., *Arthrospira platensis* and *Arthrospira maxima*
Macroalgae	*Cladophora fracta*

Table 10 Literature review LCA studies on feedstocks for biodiesel production. Adapted from Morales et al. (2015)

Year	Location	Feedstock	FU	CC	AD	AC	EP	OD	PS	ET	HT	LU	FD	WD	EN	Allocation	Methods	Source
2008	China	Soybeen	MJ	x											x	Mass	GREET and own calculations	Hu et al. (2008)
2009	Malaysia	Palm	t	x											x	N.D.	Own calculations and literature	Yee et al. (2009)
2010	Greece	Rapeseed, sunflower and soybean	Kg/ha	x		x	x	x		x		x	x			N.C.	Eco-indicator 99	Tsoutsos et al. (2010)
2011	Global	Rapeseed, sunflower and soybean	kg	x		x		x		x		x	x			N.D.	Eco-Indicator 99	Sanz Requena et al. (2011)
2012	China	Microalgae	t	x		x	x		x						x	Economic, Substitution	Research Center for Eco-Environmental Sciences (China)	Yanfen et al. (2012)
2012	US	Poultry fat	kg	x												Energy, mass	Own calculations and literature	Jørgensen et al. (2012)
2013	India	Neem and pongamia	kg	x											x	Substitution	Own calculations, literature, Net Energy Ratio (NER).	Khandelwal and Chauhan (2013)
2014	Brazil	Soybean and Palm	MJ	x	x	x	x				x					Mass, Energy	CML	Rocha et al. (2014)
2015	Indonesia	Palm and Jathropa	t	x		x	x								x	N.D.	MiLCA-JEMAI	Siregar et al. (2015)
2016	US	Jatropha and used cooking oil	t	x		x		x	x	x	x	x			x	N.D.	IMPACT 2002+	Sajid et al. (2016)
2016	Iran	Olive Pomace	MJ	x		x	x	x		x		x			x	Mass, economic and energy	IMPACT 2002+	Rajaeifar et al., (2016)

(continued)

Table 10 (continued)

Year	Location	Feedstock	FU	CC	AD	AC	EP	OD	PS	ET	HT	LU	FD	WD	EN	Allocation	Methods	Source
2016	Argentina, Spain	Rapeseed and soybean	37 MJ kg^{-1}	x		x	x	x	x		x	x	x			Mass, economic and energy	ReCiPe	Fernández-Tirado et al. (2016)
2016	Global	Microalgae	km	x								x	x			Economic, energy	ReCiPe	Gnansounou and Kenthorai Raman (2016)
2017	Italy	Linseed and Camelina	GJ and t	x		x	x	x	x	x	x		x			Economic	ILCD	Bacenetti et al. (2017)
2017	Algeria	Castor oil	t	x											x	N.C.	IMPACT 2002+, energy return-on-energy investment (EROEI)	Amouri et al. (2017)
2017	Brazil	Cotton seed	1000 kg	x	x	x	x									Mass	CML	Lima et al. (2017)
2018	US	Soybean, rapeseed and tallow	MJ	x								x				Mass, energy	GREET	Chen et al. (2018c)
2018	US	Camelina	1000 MJ	x		x	x									Energy	Impact (TRACI 2.1) and DNDC (for Soil emissions)	Tabatabaie et al. (2016)

Acronyms FU Functional Unit; *CC* Climate Change; *EN* Energy; *AD* Abiotic Depletion; *AC* Acidification; *EP* Eutrophication; *OD* Ozone Depletion; *PS* Photochemical Smog; *ET* Ecotoxicity; *HT* Human Toxicity; *LU* Land Use; *FD* Fossil Depletion; *WD* Water Depletion; *N.C.* Not considered; *N.D.* Not determined. For simplification, the impact categories in the table above are grouped: Global Warming Potential (*GWP*) comprises Climate Change (*CC*); Ecotoxicity (*ET*) and Eutrophication (*ET*) include all their derivations (terrestrial, marine, freshwater etc). Land Use (*LU*) includes also Agriculture Land Occupation (*ALO*), urban land occupation (*ULO*) and Natural Land Transformation (*NLT*). Moreover, the category "energy" (*EN*) includes net energy balance (*NEB*); net energy ratio (*NER*), cumulative energy demand (*CED*) and non-renewable energy use (NREU) to the system boundary applied in the bioethanol production chain. For instance, if the results are analysed by quantity of biodiesel or area under cultivation.

Hu et al. (2008) displayed trade-offs between soybean biodiesel and conventional diesel. The results showed that nitrogen oxides (NO_X) emissions from soybean-based biodiesel in China are nearly 80% higher than those of conventional gasoline, despite the significant reduction (67% less) in CO_2 emissions. However, the prices of biodiesel (without subsidies) are much higher (86%) than those of conventional gasoline. In Malaysia, compared to gasoline, 38% of CO_2 emissions from biodiesel combustion were reduced using palm oil biomass (Yee et al. 2009). In addition, this study indicated a net positive energy through the use of palm biodiesel, more than twice as much as rapeseed oil.

Tsoutsos et al. (2010) compared three types of biomass for biodiesel (rapeseed, sunflower and soya) in Greece, with rapeseed being the worst raw material from an environmental point of view, due to the high use of fertilisers. However, it is the most favourable bioethanol yield. Sanz Requena et al. (2011) also compared the same three feedstocks under a general geographic perspective. The results demonstrated that among all the biodiesel production processes, the cultivation of these feedstocks presented the worst impact. In addition, the impact category most affected is land use, due to the high impact on the category of damage to ecosystem quality. An LCA study on soybean and palm biodiesel was conducted using inventory data from Brazil (Rocha et al. 2014). For both feedstocks, the agricultural process is the main contributor to environmental impacts due to the use of fertilisers and pesticides. The eutrophication impact category showed a greater impact for soybean than palm oil. This is due to the high input of nutrients.

Siregar et al. (2015) compared palm oil and jatropha-based biodiesel in Indonesia and concluded that palm oil had a higher environmental impact. The application of fertilisers and the protection of plants on both crops are primarily responsible for environmental burdens. Fernández-Tirado et al. (2016) investigated rapeseed grown in Spain and soya grown in Argentina for biodiesel consumption in Spain. Soybean biodiesel from Argentina had a better environmental performance, with seed production and fertilisation being the processes with the greatest environmental impact.

Chen et al. (2018b) analysed biodiesel production in the USA, showing that, compared to gasoline diesel, soy-based biodiesel can achieve a reduction of up to 80% in fossil energy consumption and between 66 and 72% in GHG emissions. The raw material cotton was investigated under Brazilian conditions by Lima et al. (2017). Once again, the results showed that the agricultural phase represents the greatest environmental burden, due to the application of fertilisers and pesticides. In addition, the transesterification process produces liquid effluents that are released into water bodies, increasing eutrophication levels.

The production of camelina and flax seed for biofuels in the Italian Mediterranean region was investigated, showing that the seed production process represents by far the largest environmental weight for both feedstocks, due to the diesel required for agricultural machinery (Bacenetti et al. 2017). In fact, the use of fossil fuels in the agricultural process for biofuel production is quite controversial. Tabatabaie et al. (2016) investigated camelina biomass and demonstrated the importance of soil emission in LCA studies from a regional perspective. The way how agriculture is managed will have a positive or negative impact on the environment. Using non-tillage cropping system, for instance, has lower GHG emissions than conventional tillage.

In Algeria, Castor oil biomass for biodiesel presented a positive alternative environmental pathway, showing a positive reduction in climate change (Amouri et al. 2017). Additionally, in this study, the EROI value of 2.6 demonstrated a positive energy balance. The results of the LCA study by Rajaeifar et al. (2016) comparing the use of olive pomace in Iran as biodiesel (B20 and B100 blends) with conventional diesel showed positive and negative aspects. For instance, the use of biomass-based B100 blends is a better option in terms of GHG and resource damage categories, but the worst choice in terms of human health and ecosystem quality categories. Khandelwal and Chauhan (2013) considered the use of neem and pongamia plants for biodiesel production and identified that energy use during the initial processing phase of these plants is lower than the combustion phase. The wood from these plants is also valuable to the construction industry (e.g. furniture), which is an advantage over carbon sequestration.

In the USA, the use of poultry fat as a feedstock for biodiesel showed a slight 6% reduction in GHG emissions compared to diesel fuel (Jørgensen et al. 2012). This is because poultry fat already has a market value and increasing the use of biomass for biodiesel would eliminate this biomass from its initial use. Therefore, the current user of poultry fat would have to find substitutes, which would lead to an increase in GHG emissions. Sajid et al. (2016) evaluated used cooking oil and jatropha as feedstocks for biodiesel production in the United States, showing that the total environmental impact of cooking oil use is 74% lower than that of jatropha biomass. However, as the process of converting cooking oil into biodiesel requires a number of chemicals and a great deal of energy, jatropha biodiesel is more efficient.

Microalgae-based biofuels demonstrated environmental and energy benefits in China (Yanfen et al. 2012). The major impact to produce microalgae biodiesel is associated to photochemical ozone formation. The energy use for microalgae biodiesel showed an improvement of about 40% compared to its fossil counterpart. Moreover, CO_2 is the major input source that microalgae use for growth. Therefore, the higher the microalgae yield, the lower GHG is emitted. Another study on microalgae as an alternative to biodiesel and to use its co/products for chemicals and animal feed in India was reported (Gnansounou and Kenthorai 2016). The study highlighted the importance of conducting LCA studies in the early stages of product development. Not surprisingly, the results indicated that algae-based biodiesel has a lower environmental burden in relation to land use than other biofuels. However, with current technology, algae biodiesel cannot compete with terrestrial biofuels and fossil fuels in terms of energy efficiency.

- **Energy efficiency of biodiesel production from different biomass**

As indicated in the previous section, the use of second-generation biomass for biodiesel is favourable since, in general, environmental burdens are reduced. In addition, the number of studies concerning second-generation biomass, such as Jatropha, used cooking oil and even third-generation raw materials, such as algae, has increased in the last decade, as alternatives to non-edible crops. This study will perform an energy analysis of the following feedstocks: palm, jatropha, soybean, cotton seed and poultry fat, as shown in Table 11. All agricultural activities are included, as well as harvesting and transportation. A transport distance of 30 km from the field to the processing plant was assumed. In the case of poultry fat, as it is a waste, a longer distance (50 km) to the biodiesel processing plant was considered.

The feedstock yield and biodiesel potential vary widely from one study to another as they depend on a plethora of considerations: regional and climate conditions, types of agricultural practices, types of feedstock etc. Soybean and palm fruit fresh bunch (FFB) are the two main feedstocks used for biodiesel. Palm FFB has the highest feedstock yield and soybean is the least productive. However, palm FFB takes longer to be harvested and soybean has a higher biodiesel yield than palm FFB. It is important to bear in mind that jatropha also has a high production per hectare and is a second-generation biomass, which does not compete with food/feed markets. With the exception of the assessment of poultry fat, the countries considered in the feedstock analysis are one of the main agricultural producers worldwide and these feedstocks account for a significant part of their economy. Fertilisation and plant protection have very different input loads and are considered in all agricultural activities. Irrigation is not taken into consideration in the assessed feedstocks.

The energy analysis of the different biomass for biodiesel production is shown in Fig. 10. For the calculation, the EROI methodology was used, which in this case represents the relationship between the biodiesel energy supplied and the energy produced. In the case of biodiesel, the energy considered was based on its LHV (39 MJ/kg of biodiesel) according to the literature (Mata et al. 2014).The CED did not include renewable energy for chemical and agricultural assets, as the objective is only to account for fossil energy use. Poultry fat was only counted as a waste stream from collection to the biodiesel plant.

Table 11 Biomass considered for energy analyses for biodiesel

Feedstock	Source	Location	Feedstock yield	Biodiesel yield
Palm	Siregar et al. (2015)	Indonesia	82,697 kg/ha	6.38 kg/kg BDF
Jatropha	Siregar et al. (2015)	Indonesia	9,703 kg/ha	4.5 kg/kg BDF
Soybean	Chen et al. (2018c)	United States	4,676 kg/ha	5.5 kg/kg BDF
Cotton Seed	Lima et al. (2017)	Brazil	3846 kg/ha	10 kg/kg BDF
Poultry fat	Jørgensen et al. (2012)	Global	–	1.16 kg/kg BDF

It is important to note that the results of EROI may be different if treatment and utilization phases are taken into account. Each raw material and each processing plant has different processing phases and characteristics. A very important consideration is to reflect on the potentialities and trade-offs to use alternative feedstocks for biodiesel production, such as poultry fat. Jørgensen et al. (2012) demonstrated that poultry fat has already an established market in the USA and the replacement of fossil fuels by poultry fat biodiesel will lead to a reduction in the poultry fat feedstock, which may increase the necessity to find fossil fuel alternatives.

As depicted in Fig. 10, poultry fat portrays a high energy potential. The cumulative energy demand is very low, as only the transport of poultry fat waste to the plant is taken into account, which generates a very high EROI value. Jatropha in Indonesia represents the greatest potential for biodiesel and has a relatively low energy demand in the agricultural phase. In addition, although soybeans in Indonesia have a higher yield than jatropha, they require more agricultural and chemical inputs. In addition, a large amount of palm fruit is needed to produce biodiesel.

Soybean has the lower yield among the feedstocks, but is energy efficient, resulting in an EROI more than 3 MJ/MJ. Regarding cotton seed in Brazil, its production is characterised by intensive fertiliser and pesticide application. Moreover, the technology is still not developed for cotton seed biodiesel, resulting in an "energy sink". As shown in Fig. 10, the outcomes from EROI are very different for each feedstock. This is due to different amount of energy consumption, geography, climate, agricultural and processing practices. The results from poultry fat, for instance, are driven by its very low energy demand, while cotton seed by its high chemical fertiliser and pesticide demand in agriculture.

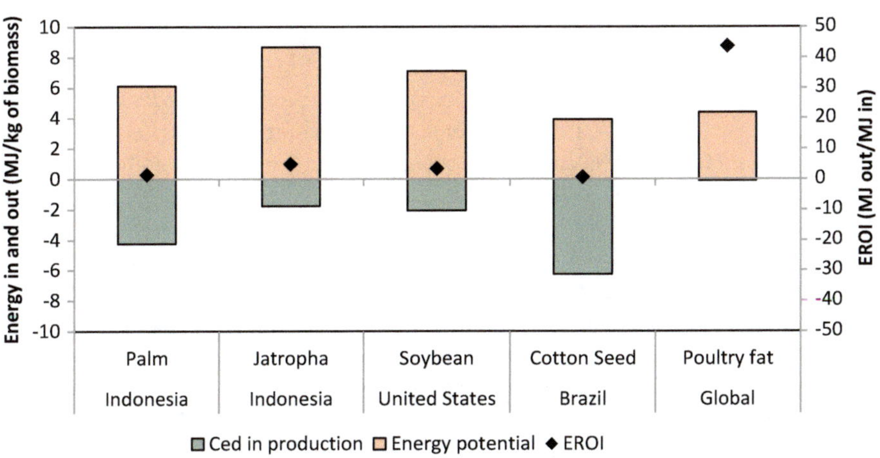

Fig. 10 Energy required and produced for each feedstock under study

4 Conclusions

One of the main challenges for the energy sector in this century will be to decouple a growing energy supply from increasing GHG emissions. Bioenergy has been considered as one of the options to address this challenge. The analysis of sustainability and energy criteria for feedstocks produced for bioenergy generation highlights the benefits of using waste streams as well as raw materials produced through logistically efficient and sustainable production systems.

This research defined the most relevant sustainability criteria in the exploitation of feedstocks for bioenergy production from a life cycle perspective. Three types of biofuels were evaluated: biogas, bioethanol and biodiesel. The comprehensive literature review on LCA studies conducted showed the diverse type of functional units found in the literature. Using a mass-oriented perspective as functional unit assumes that fuel supply is constrained by biomass production and supply. In contrast, a distance-oriented supposes biofuels supply are unlimited. An energy-driven functional unit will evaluate the quality and efficiency of this biofuel. Moreover, many studies have identified inconsistent results (from very positive to negative environmental consequences), leading to great uncertainty on this issue.

The results will be based on geographical, technological and temporal coverage and data quality. Many sustainability criteria, such as the quality of biodiversity and land use change, are very difficult to assess due to the lack of information and the difficulty of quantifying them. However, some criteria can be addressed: GHG emission intensity, consistency with legislation, resource and energy efficiency, prioritisation for waste and residues and scale of deployment (e.g. competition with food/feed stocks). The most common impact categories selected are climate change, abiotic depletion, ozone depletion, eutrophication, acidification, photochemical smog, human toxicity and fossil depletion. Land use and water depletion, although a very important impact category in relation to agriculture and forestry, were addressed in a few studies. Many studies analyse only climate change and energy as environmental indicators to understand the relationship between GHG emissions and energy. Future prospects for sustainable biofuel production include improved understanding of feedstock properties and suitable alternatives, as well as of the material and energy mass balances between production facilities. Another key issue is to assess uncertainty in LCA studies, due to the inherent variability and complexity of biofuels.

The EROI methodology was found to be suitable for comparing conventional and innovative biomass sources for bioenergy. The results vary considerably from one raw material to another, due to differences in geographical distribution, agricultural practices and energy efficiency index. In many cases, it was found that the use of by-products (e.g. maize stover for bioethanol) or waste (pig slurry for biogas) as a biomass source for biofuels could have better EROI values than first generation feedstocks due to the allocation of environmental burdens.

Biofuels currently play an important role in the economy, society and environment worldwide. However, the widespread of biofuels in the future may lead to unintended negative side-effects, such as land use and biodiversity losses. Moreover, these environmental consequences may be also increased due to the growing demand of feedstocks for the production of other products including bioplastics, biolubricants, biocomposites and biochemical, as a result of the growing European bioeconomy. However, the indicators available for quantifying these impacts are not only controversial, as already noted, but also difficult to quantify and interpret. Biofuel technology from first-generation feedstocks is now well established. However, for a sustainable future, research into the recovery of waste and by-products from raw materials must grow in order to avoid land competition as well as with the food and feed supply. Moreover, the use of abandoned land is another relevant factor in the pursuit of sustainable development of these products. LCA studies should also address not only global but regional assessments, especially when dealing with the biomass production stage. For example, aspects related to soil fertility or crop yield rely on local aspects such as soil type and climate. This study only assessed a cradle to gate analysis, from agriculture production until biomass transportation. For a global assessment, it is recommended to evaluate the energy footprint for the downstream processing stages.

Acknowledgements This research was supported by the European project STAR-ProBio (Grant Agreement Number 727740). The authors belong to the Galician Competitive Research Group GRC2013-032 and to the CRETUS Strategic Partnership (AGRUP2015/02). All these programmes are co-funded by FEDER (EU).

Glossary

BDF Biodiesel Fuel

CED Cumulative Energy Demand

CC Climate Change

CO_2 Carbon dioxide

C/N Carbon to nitrogen ratio

EIA Environmental Investigation Agency

EPA Environmental Protection Agency

EtOH Ethanol

EROI Energy Return on Investment

FFV Flexible-Fuel Vehicle

FSC Forest Stewardship Council

FBS Fuel Baseline Standard

GGL	Green Gold Label
GHG	Greenhouse gases
GWP	Global warming potential
ISCC	International Sustainability and Carbon Certification System
ITC	The International Trade Centre
LCA	Life Cycle Assessment
LHV	Low Heating Value
LUC	Land Use Change
NER	Net energy Ratio
OFMSW	Organic Fraction of Municipal Solid Waste
RA	Rainforest Alliance
RED	Renewable Energy Directive
RSB	Roundtable on Sustainable Biomaterials
RSPO	Roundtable on Sustainable Palm Oil
RTRS	Round Table Responsible Soy
SAN	Sustainable Agriculture Network

References

2BSvs. (2016). 2BS biomass biofuels sustainability voluntary scheme: Scheme description.

Al Seadi, T., Drosg, B., & Fuchs, W. (2013). Biogas digestate quality and utilization. In A. Wellinger, J. Murphy, D. Baxter (Eds.), *The biogas handbook* (p. 476). Science, Production and Applications.

Abas, N., Kalair, A., & Khan, N. (2015). Review of fossil fuels and future energy technologies. *Futures, 69,* 31–49. https://doi.org/10.1016/j.futures.2015.03.003.

Alexiades, A., Kendall, A., Winans, K. S., & Kaffka, S. R. (2018). Sugar beet ethanol (*Beta vulgaris L.*): A promising low-carbon pathway for ethanol production in California. *Journal of Cleaner Production, 172,* 3907–3917. https://doi.org/10.1016/j.jclepro.2017.05.059.

Allen, B., Baldock, D., Nanni, S., & Bowyer, C. (2016). Sustainability criteria for biofuels made from land and non-land based feedstocks. *Institute for Environmental European Policy.*

Akbulut, A. (2012). Techno-economic analysis of electricity and heat generation from farm-scale biogas plant: Çiçekdağı case study. *Energy, 44*, 381–390, Elsevier. https://doi.org/10.1016/j.energy.2012.06.017.

Amouri, M., Mohellebi, F., Zaïd, T. A., & Aziza, M. (2017). Sustainability assessment of Ricinus communis biodiesel using LCA approach. *Clean Technologies and Environmental Policy, 19,* 749–760. https://doi.org/10.1007/s10098-016-1262-4.

Anastopoulos, G., Zannikou, Y., Stournas, S., & Kalligeros, S. (2009). Transesterification of vegetable oils with ethanol and characterization of the key fuel properties of ethyl esters. *Energies, 2,* 362–376. https://doi.org/10.3390/en20200362.

Arifeen, N., Kookos, I. K., Wang, R., Koutinas, A. A., & Webb, C. (2009). Development of novel wheat biorefining: Effect of gluten extraction from wheat on bioethanol production. *Biochemical Engineering Journal, 43,* 113–121. https://doi.org/10.1016/j.bej.2008.09.005.

Arvidsson, R., Fransson, K., Fröling, M., Svanström, M., & Molander, S. (2012). Energy use indicators in energy and life cycle assessments of biofuels: Review and recommendations. *Journal of Cleaner Production, 31,* 54–61. https://doi.org/10.1016/j.jclepro.2012.03.001.

Bacenetti, J., Restuccia, A., Schillaci, G., & Failla, S. (2017). Biodiesel production from unconventional oilseed crops (*Linum usitatissimum L.* and *Camelina sativa L.*) in Mediterranean conditions: Environmental sustainability assessment. *Renewable Energy, 112,* 444–456. https://doi.org/10.1016/j.renene.2017.05.044.

Bai, Y., Luo, L., & Van Der Voet, E. (2010). Life cycle assessment of switchgrass-derived ethanol as transport fuel. *International Journal of Life Cycle Assessment, 15,* 468–477. https://doi.org/10.1007/s11367-010-0177-2.

Bansal, A., Illukpitiya, P., Tegegne, F., & Singh, S. P. (2016). Energy efficiency of ethanol production from cellulosic feedstock. *Renewable and Sustainable Energy Reviews, 58,* 141–146. https://doi.org/10.1016/j.rser.2015.12.122.

Beurskens, L. W. M., & Hekkenberg, M. (2011). *Renewable Energy Projections as Published in the National Renewable Energy Action Plans of the European Member States Covering all 27 EU Member States* (p. 244). Netherlands: Energy Research Centre.

Bithas, K., & Kalimeris, P. (2016). A brief history of energy use in human societies. In *Revisiting the energy-development link* (pp. 5–11). https://doi.org/10.1007/978-3-319-20732-2.

Bonn, M., Heitmann, N., Reichert, G., & Voßwinkel, J.S. (2015). EU Climate and Energy Policy 2030 Comments on an Evolving Framework.

Borines, M. G., de Leon, R. L., & Cuello, J. L. (2013). Bioethanol production from the macroalgae Sargassum spp. *Bioresource Technology, 138,* 22–29. https://doi.org/10.1016/j.biortech.2013.03.108.

Borrion, A.L., McManus, M.C., & Hammond, G.P. (2012). Environmental life cycle assessment of bioethanol production from wheat straw. *Biomass and Bioenergy*, *47*, 9–19. https://doi.org/10.1016/j.biombioe.2012.10.017.

Bosch, R., Van De Pol, M., & Philp, J. (2015). Policy: Define biomass sustainability. *Nature, 523,* 526–527. https://doi.org/10.1038/523526a.

Buchholz, T., Luzadis, V. A., & Volk, T. A. (2009). Sustainability criteria for bioenergy systems: results from an expert survey. *Journal of Cleaner Production, 17,* S86–S98. https://doi.org/10.1016/j.jclepro.2009.04.015.

Bühle, L., Hensgen, F., Donnison, I., Heinsoo, K., & Wachendorf, M. (2012). Life cycle assessment of the integrated generation of solid fuel and biogas from biomass (IFBB) in comparison to different energy recovery, animal-based and non-refining management systems. *Bioresource Technology, 111,* 230–239. https://doi.org/10.1016/j.biortech.2012.05.077.

Bühle, L., Stülpnagel, R., Wachendorf, M. (2011). Comparative life cycle assessment of the integrated generation of solid fuel and biogas from biomass (IFBB) and whole crop digestion (WCD) in Germany. *Biomass and Bioenergy, 35*, 363–373. https://doi.org/10.1016/j.biombioe.2010.08.056.

Carneiro, M. L. N. M., Pradelle, F., Braga, S. L., Gomes, M. S. P., Martins, A. R. F. A., Turkovics, F., et al. (2017). Potential of biofuels from algae: Comparison with fossil fuels, ethanol and biodiesel in Europe and Brazil through life cycle assessment (LCA). *Renewable and Sustainable Energy Reviews, 73,* 632–653. https://doi.org/10.1016/j.rser.2017.01.152.

Chen, S., Xu, Z., Li, X., Yu, J., Cai, M., & Jin, M. (2018a). Integrated bioethanol production from mixtures of corn and corn stover. *Bioresource Technology, 258,* 18–25. https://doi.org/10.1016/j.biortech.2018.02.125.

Chen, J., Tyagi, R. D., Li, J., Zhang, X., Drogui, P., & Sun, F. (2018b). Economic assessment of biodiesel production from wastewater sludge. *Bioresource Technology, 253,* 41–48. https://doi.org/10.1016/j.biortech.2018.01.016.

Chen, R., Qin, Z., Han, J., Wang, M., Taheripour, F., Tyner, W., et al. (2018c). Life cycle energy and greenhouse gas emission effects of biodiesel in the United States with induced land use change impacts. *Bioresource Technology, 251,* 249–258. https://doi.org/10.1016/j.biortech.2017.12.031.

Cherubini, F., Bird, N. D., Cowie, A., Jungmeier, G., Schlamadinger, B., & Woess-Gallasch, S. (2009). Energy- and greenhouse gas-based LCA of biofuel and bioenergy systems: Key issues, ranges and recommendations. *Resources, Conservation and Recycling, 53,* 434–447. https://doi.org/10.1016/j.resconrec.2009.03.013.

Cherubini, F., & Strømman, A. H. (2011). Life cycle assessment of bioenergy systems: State of the art and future challenges. *Bioresource Technology, 102,* 437–451. https://doi.org/10.1016/j.biortech.2010.08.010.

Chia, S. R., Ong, H. C., Chew, K. W., Show, P. L., Phang, S. M., & Ling, T. C. et al. (2017). Sustainable approaches for algae utilisation in bioenergy production. *Renewable Energy.* https://doi.org/10.1016/j.renene.2017.04.001.

Da Costa-Gomez, C. (2013). Biogas as an energy option: An overview. In A. Wellinger, J. Murphy, D. Baxter (Eds.), *Biogas handbook* (p. 476). Science, Production and Applications.

de Azevedo, A., Fornasier, F., da Silva Szarblewski, M., de Souza Schneider, R. D. C., Hoeltz, M., & de Souza, D. (2017). Life cycle assessment of bioethanol production from cattle manure. *Journal of Cleaner Production, 162*, 1021–1030. https://doi.org/10.1016/j.jclepro.2017.06.141.

Dressler, D., Loewen, A. & Nelles, M. (2012). Life cycle assessment of the supply and use of bioenergy: Impact of regional factors on biogas production. *International Journal Life Cycle Assessment, 17*, 1104–1015. https://doi.org/10.1007/s11367-012-0424-9.

De Vries, J. W., Vinken, T. M. W. J., Hamelin, L., & De Boer, I. J. M. (2012). Comparing environmental consequences of anaerobic mono- and co-digestion of pig manure to produce bio-energy -A life cycle perspective. *Bioresource Technology, 125,* 239–248. https://doi.org/10.1016/j.biortech.2012.08.124.

Debnath, D., Whistance, J., Thompson, W., & Binfield, J. (2017). Complement or substitute: Ethanol's uncertain relationship with gasoline under alternative petroleum price and policy scenarios. *Applied Energy, 191,* 385–397. https://doi.org/10.1016/j.apenergy.2017.01.028.

Deloitte. (2014). Opportunities for the fermentation-based chemical industry: An analysis of the market potential and competitiveness of North-West Europe, 1–50. https://doi.org/10.2767/39856.

Duraisam, R., Salelgn, K., & Berekete, A. K. (2017). Production of beet sugar and bio-ethanol from sugar beet and it bagasse: A Review. *International Journal of Engineering Trends Technology, 43*, 222–233. https://doi.org/10.14445/22315381/IJETT-V43P237.

Ebner, J., Babbitt, C., Winer, M., Hilton, B., & Williamson, A. (2014). Life cycle greenhouse gas (GHG) impacts of a novel process for converting food waste to ethanol and co-products. *Applied Energy, 130,* 86–93. https://doi.org/10.1016/j.apenergy.2014.04.099.

Efe, Ş., Ceviz, M. A., & Temur, H. (2018). Comparative engine characteristics of biodiesels from hazelnut, corn, soybean, canola and sunflower oils on DI diesel engine. *Renewable Energy, 119,* 142–151. https://doi.org/10.1016/j.renene.2017.12.011.

Einarsson, R., & Persson, U. M. (2017). Analyzing key constraints to biogas production from crop residues and manure in the EU—A spatially explicit model. *PLoS ONE, 12,* 1–23. https://doi.org/10.1371/journal.pone.0171001.

EPA. (2010). Regulation of fuels and fuel additives: Changes to renewable fuel standard program. *Federal Register.* https://doi.org/10.1080/19388078109557636.

European Commission. (2014). State of play on the sustainability of solid and gaseous biomass used for electricity, heating and cooling in the EU. Brussels, SWD 259 final.

European Parliament. (2009). Directive 2009/28/EC of the European Parliament and of the Council of 23 April 2009. *Official Journal of European Union, 140*, 16–62. https://doi.org/10.3000/17252555.L_2009.140.eng.

Eurostat. (2017). Energy balance sheets. https://doi.org/10.2785/032728.

Fantin, V., Giuliano, A., Manfredi, M., Ottaviano, G., Stefanova, M., & Masoni, P. (2015). Environmental assessment of electricity generation from an Italian anaerobic digestion plant. *Biomass and Bioenergy, 83*, 422–435.https://doi.org/10.1016/j.biombioe.2015.10.015.

Feiz, R., & Ammenberg, J. (2017). Assessment of feedstocks for biogas production, part I—A multi-criteria approach. *Resources, Conservation and Recycling, 122,* 373–387. https://doi.org/10.1016/j.resconrec.2017.01.019.

Fernández-Tirado, F., Parra-López, C., & Romero-Gámez, M. (2016). Life cycle assessment of biodiesel in Spain: Comparing the environmental sustainability of Spanish production versus Argentinean imports. *Energy for Sustainable Development, 33,* 36–52. https://doi.org/10.1016/j.esd.2016.04.005.

Forte, A., Zucaro, A., Fagnano, M., & Fierro, A. (2017). Potential environmental impact of bioethanol production chain from fiber sorghum to be used in passenger cars. *Science of the Total Environment, 598,* 365–376. https://doi.org/10.1016/j.scitotenv.2017.03.244.

Forte, A., Zucaro, A., Faugno, S., Basosi, R., & Fierro, A. (2018). Carbon footprint and fossil energy consumption of bio-ethanol fuel production from *Arundo donax L.* crops on marginal lands of Southern Italy. *Energy*. https://doi.org/10.1016/j.energy.2018.02.030.

Foteinis, S., Kouloumpis, V., & Tsoutsos, T. (2011). Life cycle analysis for bioethanol production from sugar beet crops in Greece. *Energy Policy, 39,* 4834–4841. https://doi.org/10.1016/j.enpol.2011.06.036.

Frischknecht, R., Wyss, F., Büsser Knöpfel, S., Lützkendorf, T., & Balouktsi, M. (2015). Cumulative energy demand in LCA: The energy harvested approach. *International Journal of Life Cycle Assessment, 20,* 957–969. https://doi.org/10.1007/s11367-015-0897-4.

Fritsche, U.R., Hennenberg, K., Hünecke, K., & Wiegmann, K. (2012). Sustainable bioenergy: Key Criteria and Indicators Final D 4. 1 Delivery of the project funded by with inputs from Klaus Hennenberg, Katja Hünecke.

Fritsche, U. R., & Iriarte, L. (2014). Sustainability criteria and indicators for the bio-based economy in Europe: State of discussion and way forward. *Energies, 7,* 6825–6836. https://doi.org/10.3390/en7116825.

FSC. (2017). How does FSC ensure climate benefits and prevent negative impacts from woody biomass energy.

Gissén, C., Prade, T., Kreuger, E., Nges, I. A., Rosenqvist, H., Svensson, S. E., et al. (2014). Comparing energy crops for biogas production—Yields, energy input and costs in cultivation using digestate and mineral fertilisation. *Biomass and Bioenergy, 64,* 199–210. https://doi.org/10.1016/j.biombioe.2014.03.061.

Giuffrè, A. M., Capocasale, M., Zappia, C., Sicari, V., Pellicanò, T. M., Poiana, M., et al. (2016). Tomato seed oil for biodiesel production. *European Journal of Lipid Science and Technology, 118,* 640–650. https://doi.org/10.1002/ejlt.201500002.

Giuntoli, J., Boulamanti, A. K., Corrado, S., Motegh, M., Agostini, A., & Baxter, D. (2013). Environmental impacts of future bioenergy pathways: The case of electricity from wheat straw bales and pellets. *GCB Bioenergy, 5,* 497–512. https://doi.org/10.1111/gcbb.12012.

Giwa, A. (2017). Comparative cradle-to-grave life cycle assessment of biogas production from marine algae and cattle manure biorefineries. *Bioresource Technology, 244,* 1470–1479. https://doi.org/10.1016/j.biortech.2017.05.143.

Gnansounou, E., & Kenthorai Raman, J. (2016). Life cycle assessment of algae biodiesel and its co-products. *Applied Energy, 161,* 300–308. https://doi.org/10.1016/j.apenergy.2015.10.043.

González-García, S., Bacenetti, J., Negri, M., Fiala, M., & Arroja, L. (2013). Comparative environmental performance of three different annual energy crops for biogas production in Northern Italy. *Journal of Cleaner Production, 43,* 71–83. https://doi.org/10.1016/j.jclepro.2012.12.017.

González-García, S., Moreira, M. T., & Feijoo, G. (2010). Comparative environmental performance of lignocellulosic ethanol from different feedstocks. *Renewable and Sustainable Energy Reviews, 14,* 2077–2085. https://doi.org/10.1016/j.rser.2010.03.035.

Guerrero, A. B., & Muñoz, E. (2018). Life cycle assessment of second generation ethanol derived from banana agricultural waste: Environmental impacts and energy balance. *Journal of Cleaner Production, 174,* 710–717. https://doi.org/10.1016/j.jclepro.2017.10.298.

Guo, M., Song, W., & Buhain, J. (2015). Bioenergy and biofuels: History, status, and perspective. *Renewable and Sustainable Energy Reviews, 42,* 712–725. https://doi.org/10.1016/j.rser.2014.10.013.

Hall, C. A. S., Lambert, J. G., & Balogh, S. B. (2014). EROI of different fuels and the implications for society. *Energy Policy, 64,* 141–152. https://doi.org/10.1016/j.enpol.2013.05.049.

Hanif, M., Mahlia, T. M. I., Aditiya, H. B., & Abu Bakar, M. S. (2017). Energy and environmental assessments of bioethanol production from Sri Kanji 1 cassava in Malaysia. *Biofuel Research Journal, 4,* 537–544. https://doi.org/10.18331/BRJ2017.4.1.3.

Ho, S. H., Wong, Y. D., & Chang, V. W. C. (2014). Evaluating the potential of biodiesel (via recycled cooking oil) use in Singapore, an urban city. *Resources, Conservation and Recycling, 91,* 117–124. https://doi.org/10.1016/j.resconrec.2014.08.003.

Holm-Nielsen, J. B., Al Seadi, T., & Oleskowicz-Popiel, P. (2009). The future of anaerobic digestion and biogas utilization. *Bioresource Technology, 100,* 5478–5484. https://doi.org/10.1016/j.biortech.2008.12.046.

Hossain, M. N. Bin, Basu, J. K., & Mamun, M. (2015). The production of ethanol from micro-algae spirulina. *Procedia Eng., 105,* 733–738. https://doi.org/10.1016/j.proeng.2015.05.064.

Hu, Z., Tan, P., Yan, X., & Lou, D. (2008). Life cycle energy, environment and economic assessment of soybean-based biodiesel as an alternative automotive fuel in China. *Energy, 33,* 1654–1658. https://doi.org/10.1016/j.energy.2008.06.004.

IEA. (2017a). *CO_2 emissions from fuel combustion 2017—Highlights*. International Energy Agency. https://doi.org/10.1787/co2_fuel-2017-en.

IEA. (2017b). *Technology roadmap: Delivering sustainable bioenergy*. International Energy Agency.

Immerzeel, D. J., Verweij, P. A., van der Hilst, F., & Faaij, A. P. C. (2014). Biodiversity impacts of bioenergy crop production: A state-of-the-art review. *GCB Bioenergy, 6,* 183–209. https://doi.org/10.1111/gcbb.12067.

Isikgor, F. H., & Becer, C. R. (2015). Lignocellulosic biomass: A sustainable platform for the production of bio-based chemicals and polymers. *Polymer Chemistry, 6,* 4497–4559. https://doi.org/10.1039/C5PY00263J.

ISO Standard. (2011). *Sustainability criteria for bioenergy*.

Jacobs, A., Auburger, S., Bahrs, E., Brauer-Siebrecht, W., Christen, O., Götze, P., et al. (2017). Greenhouse gas emission of biogas production out of silage maize and sugar beet—An assessment along the entire production chain. *Applied Energy, 190,* 114–121. https://doi.org/10.1016/j.apenergy.2016.12.117.

Jørgensen, A., Bikker, P., & Herrmann, I. T. (2012). Assessing the greenhouse gas emissions from poultry fat biodiesel. *Journal of Cleaner Production, 24,* 85–91. https://doi.org/10.1016/j.jclepro.2011.11.011.

Kazamia, E., & Smith, A. G. (2014). Assessing the environmental sustainability of biofuels. *Trends in Plant Science, 19,* 615–618. https://doi.org/10.1016/j.tplants.2014.08.001.

Khan, S., Siddique, R., Sajjad, W., Nabi, G., Hayat, K. M., Duan, P., et al. (2017). Biodiesel production from algae to overcome the energy crisis. *HAYATI Journal of Biosciences, 24,* 163–167. https://doi.org/10.1016/j.hjb.2017.10.003.

Khandelwal, S., & Chauhan, R. Y. (2013). Life cycle assessment of Neem and Karanja biodiesel: An overview. *International Journal of ChemTech Research, 5,* 659–665.

Kligerman, D. C., & Bouwer, E. J. (2015). Prospects for biodiesel production from algae-based wastewater treatment in Brazil: A review. *Renewable and Sustainable Energy Reviews, 52,* 1834–1846. https://doi.org/10.1016/j.rser.2015.08.030.

Knoll, K., West, B., Huff, S., Thomas, J., Orban, J., & Cooper, C. (2009). Effects of Mid-Level Ethanol Blends on Conventional Vehicle Emissions. *SAE Technical Paper, 4970,* 2–4. https://doi.org/10.4271/2009-01-2723.

Lelieveld, J., Evans, J. S., Fnais, M., Giannadaki, D., & Pozzer, A. (2015). The contribution of outdoor air pollution sources to premature mortality on a global scale. *Nature, 525,* 367–371. https://doi.org/10.1038/nature15371.

Lendle, A., & Schaus, M. (2010). Sustainability criteria in the EU renewable energy directive: Consistent with WTO Rules? *International Centre for Trade and Sustainable Development, 16.*

Lijó, L., González-García, S., Bacenetti, J., & Moreira, M. T. (2017). The environmental effect of substituting energy crops for food waste as feedstock for biogas production. *Energy, 137,* 1130–1143. https://doi.org/10.1016/j.energy.2017.04.137.

Lijó, L., González-García, S., Bacenetti, J., Negri, M., Fiala, M., Feijoo, G., et al. (2015). Environmental assessment of farm-scaled anaerobic co-digestion for bioenergy production. *Waste Management, 41,* 50–59. https://doi.org/10.1016/j.wasman.2015.03.043.

Lima, Â. M. F., Torres, E. A., Kiperstok, A., & de Freitas Moreira Santos, G. (2017). Environmental impacts of the biodiesel production chain of cotton seed in Bahia, Brazil. *Clean Technology Environment Policy, 19,* 1523–1534. https://doi.org/10.1007/s10098-017-1347-8.

Luo, L., Van Der Voet, E., Huppes, G., & Udo De Haes, H. A. (2009). Allocation issues in LCA methodology: A case study of corn stover-based fuel ethanol. *International Journal of Life Cycle Assessment, 14,* 529–539. https://doi.org/10.1007/s11367-009-0112-6.

Lynd, L. R., Liang, X., Biddy, M. J., Allee, A., Cai, H., Foust, T., et al. (2017). Cellulosic ethanol: Status and innovation. *Current Opinion in Biotechnology, 45,* 202–211. https://doi.org/10.1016/j.copbio.2017.03.008.

Machado, K. S., Seleme, R., Maceno, M. M., & Zattar, I. C. (2017). Carbon footprint in the ethanol feedstocks cultivation—Agricultural CO2 emission assessment. *Agricultural Systems, 157,* 140–145. https://doi.org/10.1016/j.agsy.2017.07.015.

Markeviius, A., Katinas, V., Perednis, E., & Tamaauskien, M. (2010). Trends and sustainability criteria of the production and use of liquid biofuels. *Renewable and Sustainable Energy Reviews, 14,* 3226–3231. https://doi.org/10.1016/j.rser.2010.07.015.

Mata, T. M., Mendes, A. M., Caetano, N. S., & Martins, A. A. (2014). Properties and sustainability of biodiesel from animal fats and fish oil. *Chemical Engineering Transactions, 38,* 175–180. https://doi.org/10.3303/CET1438030.

Mela, G., & Canali, G. (2014). How distorting policies can affect energy efficiency and sustainability: The case of biogas production in the Po Valley (Italy). *AgBioForum, 16,* 194–206.

Meyer, M. A., & Priess, J. A. (2014). Indicators of bioenergy-related certification schemes - An analysis of the quality and comprehensiveness for assessing local/regional environmental impacts. *Biomass and Bioenergy, 65,* 151–169. https://doi.org/10.1016/j.biombioe.2014.03.041.

Mezzullo, W. G., McManus, M. C., & Hammond, G. P. (2013). Life cycle assessment of a small-scale anaerobic digestion plant from cattle waste. *Applied Energy, 102,* 657–664. https://doi.org/10.1016/j.apenergy.2012.08.008.

Mitchell, R., Vogel, K. P., & Uden, D. R. (2012). The feasibility of switchgrass for biofuel production. *Biofuels, 3,* 47–59. https://doi.org/10.4155/bfs.11.153.

Morales, M., Quintero, J., Conejeros, R., & Aroca, G. (2015). Life cycle assessment of lignocellulosic bioethanol: Environmental impacts and energy balance. *Renewable and Sustainable Energy Reviews, 42,* 1349–1361. https://doi.org/10.1016/J.RSER.2014.10.097.

Muñoz, I., Flury, K., Jungbluth, N., Rigarlsford, G., I Canals, L.M., & King, H. (2014). Life cycle assessment of bio-based ethanol produced from different agricultural feedstocks. *International Journal of Life Cycle Assessment*, *19*, 109–119. https://doi.org/10.1007/s11367-013-0613-1.

Nakanishi, S. C., Nascimento, V. M., Rabelo, S. C., Sampaio, I. L. M., Junqueira, T. L., & Rocha, G. J. M. (2018). Comparative material balances and preliminary technical analysis of the pilot scale sugarcane bagasse alkaline pretreatment to 2G ethanol production. *Industrial Crops and Products, 120,* 187–197. https://doi.org/10.1016/j.indcrop.2018.04.064.

Negri, M., Bacenetti, J., Brambilla, M., Manfredini, A., Cantore, A., & Bocchi, S. (2014). Biomethane production from different crop systems of cereals in Northern Italy. *Biomass and Bioenergy, 63,* 321–329. https://doi.org/10.1016/j.biombioe.2014.01.041.

Negri, M., Bacenetti, J., Fiala, M., & Bocchi, S. (2016). Evaluation of anaerobic degradation, biogas and digestate production of cereal silages using nylon-bags. *Bioresource Technology, 209*, 40–49. https://doi.org/10.1016/j.biortech.2016.02.101.

Neupane, B., Halog, A., & Dhungel, S. (2011). Attributional life cycle assessment of woodchips for bioethanol production. *Journal of Cleaner Production, 19,* 733–741. https://doi.org/10.1016/j.jclepro.2010.12.002.

Nicoletti, G., Arcuri, N., Nicoletti, G., & Bruno, R. (2015). A technical and environmental comparison between hydrogen and some fossil fuels. *Energy Conversion and Management, 89,* 205–213. https://doi.org/10.1016/j.enconman.2014.09.057.

Norton-Brandão, D., Scherrenberg, S. M., & van Lier, J. B. (2013). Reclamation of used urban waters for irrigation purposes—A review of treatment technologies. *Journal of Environmental Management, 122,* 85–98. https://doi.org/10.1016/j.jenvman.2013.03.012.

NTA8080. (2011). Netherlands technical agreement: Certification scheme for sustainably produced biomass for energy purposes (Version 1.4).

Oh, P. P., Lau, H. L. N., Chen, J., Chong, M. F., & Choo, Y. M. (2012). A review on conventional technologies and emerging process intensification (PI) methods for biodiesel production. *Renewable and Sustainable Energy Reviews, 16,* 5131–5145. https://doi.org/10.1016/j.rser.2012.05.014.

Piloto-Rodríguez, R., Sánchez-Borroto, Y., Melo-Espinosa, E. A., & Verhelst, S. (2017). Assessment of diesel engine performance when fueled with biodiesel from algae and microalgae: An overview. *Renewable and Sustainable Energy Reviews, 69,* 833–842. https://doi.org/10.1016/j.rser.2016.11.015.

Poeschl, M., Ward, S., & Owende, P. (2012a). Environmental impacts of biogas deployment—Part I: Life cycle inventory for evaluation of production process emissions to air. *Journal of Cleaner Production, 24,* 168–183. https://doi.org/10.1016/j.jclepro.2011.10.039.

Poeschl, M., Ward, S., & Owende, P. (2012b). Environmental impacts of biogas deployment—Part II: Life cycle assessment of multiple production and utilization pathways. *Journal of Cleaner Production, 24,* 184–201. https://doi.org/10.1016/j.jclepro.2011.10.030.

Rainforest Alliance. (2015). SAN/Rainforest Alliance Impacts Report 2015 [WWW Document]. https://www.rainforest-alliance.org/impact-studies/impacts-report-2015.

Rajaeifar, M. A., Akram, A., Ghobadian, B., Rafiee, S., Heijungs, R., & Tabatabaei, M. (2016). Environmental impact assessment of olive pomace oil biodiesel production and consumption: A comparative lifecycle assessment. *Energy, 106,* 87–102. https://doi.org/10.1016/j.energy.2016.03.010.

Rehan, M., Gardy, J., Demirbas, A., Rashid, U., Budzianowski, W. M., Pant, D., et al. (2018). Waste to biodiesel: A preliminary assessment for Saudi Arabia. *Bioresource Technology, 250,* 17–25. https://doi.org/10.1016/j.biortech.2017.11.024.

Rocha, M. H., Capaz, R. S., Lora, E. E. S., Nogueira, L. A. H., Leme, M. M. V., Renó, M. L. G., et al. (2014). Life cycle assessment (LCA) for biofuels in Brazilian conditions: A meta-analysis. *Renewable and Sustainable Energy Reviews, 37,* 435–459. https://doi.org/10.1016/j.rser.2014.05.036.

RSB. (2016). Roundtable on sustainable biomaterials: Principles and criteria.

Sadik, M. W., & Halema, A. A. (2014). Production of Ethanol from molasses and whey permeate using yeasts and bacterial strains. *International Journal of Current Microbiology and Applied Sciences, 3,* 804–818.

Sajid, Z., Khan, F., & Zhang, Y. (2016). Process simulation and life cycle analysis of biodiesel production. *Renewable Energy, 85,* 945–952. https://doi.org/10.1016/j.renene.2015.07.046.

Sanz Requena, J. F., Guimaraes, A. C., Quirós Alpera, S., Relea Gangas, E., Hernandez-Navarro, S., Navas Gracia, L. M., et al. (2011). Life Cycle Assessment (LCA) of the biofuel production process from sunflower oil, rapeseed oil and soybean oil. *Fuel Processing Technology, 92,* 190–199. https://doi.org/10.1016/j.fuproc.2010.03.004.

Scarlat, N., Dallemand, J. F., Monforti-Ferrario, F., & Nita, V. (2015). The role of biomass and bioenergy in a future bioeconomy: Policies and facts. *Environmental Development, 15,* 3–34. https://doi.org/10.1016/j.envdev.2015.03.006.

Sebastião, D., Gonçalves, M. S., Marques, S., Fonseca, C., Gírio, F., Oliveira, A. C., et al. (2016). Life cycle assessment of advanced bioethanol production from pulp and paper sludge. *Bioresource Technology, 208,* 100–109. https://doi.org/10.1016/j.biortech.2016.02.049.

Seber, G., Malina, R., Pearlson, M. N., Olcay, H., Hileman, J. I., & Barrett, S. R. H. (2014). Environmental and economic assessment of producing hydroprocessed jet and diesel fuel from waste oils and tallow. *Biomass and Bioenergy, 67,* 108–118. https://doi.org/10.1016/j.biombioe.2014.04.024.

Siregar, K., Tambunan, A. H., Irwanto, A. K., Wirawan, S. S., & Araki, T. (2015). A Comparison of Life Cycle Assessment on Oil Palm (*Elaeis guineensis Jacq.*) and Physic Nut (*Jatropha curcas Linn.*) as Feedstock for Biodiesel Production in Indonesia. *Energy Procedia, 65,* 170–179. https://doi.org/10.1016/j.egypro.2015.01.054.

Sivaramakrishnan, R., & Incharoensakdi, A. (2018). Utilization of microalgae feedstock for concomitant production of bioethanol and biodiesel. *Fuel, 217,* 458–466. https://doi.org/10.1016/j.fuel.2017.12.119.

Soam, S., Kumar, R., Gupta, R. P., Sharma, P. K., Tuli, D. K., & Das, B. (2015). Life cycle assessment of fuel ethanol from sugarcane molasses in northern and western India and its impact on Indian biofuel programme. *Energy, 83,* 307–315. https://doi.org/10.1016/j.energy.2015.02.025.

Song, C., Qiu, Y., Liu, Q., Ji, N., Zhao, Y., & Kitamura, Y. et al. 2018. Process intensification of cellulosic ethanol production by waste heat integration. *Chemistry Engineering Research and Design*. https://doi.org/10.1016/j.cherd.2018.01.016.

Sonnemann, G., Gemechu, E. D., Adibi, N., De Bruille, V., & Bulle, C. (2015). From a critical review to a conceptual framework for integrating the criticality of resources into life cycle sustainability assessment. *Journal of Cleaner Production, 94,* 20–34. https://doi.org/10.1016/j.jclepro.2015.01.082.

Tabatabaie, S. M. H., Tahami, H., & Murthy, G. S. (2016). A regional life cycle assessment and economic analysis of camelina biodiesel production in the Pacific Northwestern US. *Journal of Cleaner Production, 172,* 2389–2400. https://doi.org/10.1016/j.jclepro.2017.11.172.

The German Federal Government. (2012). *Biorefineries roadmap*. Berlin.

Timilsina, G.R., & Shrestha, A. (2010). Biofuels markets, targets and impacts. *World*, 1–47. https://doi.org/10.1596/1813-9450-5364.

Tsiropoulos, I., Faaij, A. P. C., Seabra, J. E. A., Lundquist, L., Schenker, U., Briois, J. F., et al. (2014). Life cycle assessment of sugarcane ethanol production in India in comparison to Brazil. *International Journal of Life Cycle Assessment, 19,* 1049–1067. https://doi.org/10.1007/s11367-014-0714-5.

Tsoutsos, T., Kouloumpis, V., Zafiris, T., & Foteinis, S. (2010). Life Cycle Assessment for biodiesel production under Greek climate conditions. *Journal of Cleaner Production, 18,* 328–335. https://doi.org/10.1016/j.jclepro.2009.11.002.

Tumuluru, J. S. (2015). Comparison of chemical composition and energy property of torrefied switchgrass and corn stover front. *Energy Research, 3,* 1–11. https://doi.org/10.3389/fenrg.2015.00046.

Tutt, M., Kikas, T., & Olt, J. (2012). Influence of different pretreatment methods on bioethanol production from wheat straw. *Agronomy Research Biosystems Engineeering*, 269–276.

Van Dam, J., & Junginger, M. (2011). Striving to further harmonization of sustainability criteria for bioenergy in Europe: Recommendations from a stakeholder questionnaire. *Energy Policy, 39,* 4051–4066. https://doi.org/10.1016/j.enpol.2011.03.022.

Vega, G. C. C., ten Hoeve, M., Birkved, M., Sommer, S. G., & Bruun, S. (2014). Choosing co-substrates to supplement biogas production from animal slurry—A life cycle assessment of the environmental consequences. *Bioresource Technology, 171,* 410–420. https://doi.org/10.1016/j.biortech.2014.08.099.

Wang, L., Littlewood, J., & Murphy, R. J. (2013). Environmental sustainability of bioethanol production from wheat straw in the UK. *Renewable and Sustainable Energy Reviews, 28,* 715–725. https://doi.org/10.1016/j.rser.2013.08.031.

Wang, M., Han, J., Dunn, J. B., Cai, H., & Elgowainy, A. (2012). Well-to-wheels energy use and greenhouse gas emissions of ethanol from corn, sugarcane and cellulosic biomass for US use. *Environoment Research Letters, 7.* https://doi.org/10.1088/1748-9326/7/4/045905.

Whiting, A., & Azapagic, A. (2014). Life cycle environmental impacts of generating electricity and heat from biogas produced by anaerobic digestion. *Energy, 70,* 181–193. https://doi.org/10.1016/j.energy.2014.03.103.

World Energy Council. (2016). World Energy Resources, @WECouncil. http://www.worldenergy.org/wp-content/uploads/2013/09/Complete_WER_2013_Survey.pdf.

Yanfen, L., Zehao, H., & Xiaoqian, M. (2012). Energy analysis and environmental impacts of microalgal biodiesel in China. *Energy Policy, 45,* 142–151. https://doi.org/10.1016/j.enpol.2012.02.007.

Yee, K. F., Tan, K. T., Abdullah, A. Z., & Lee, K. T. (2009). Life cycle assessment of palm biodiesel: Revealing facts and benefits for sustainability. *Applied Energy, 86,* S189–S196. https://doi.org/10.1016/j.apenergy.2009.04.014.

Zheng, L., Hou, Y., Li, W., Yang, S., Li, Q., & Yu, Z. (2012). Biodiesel production from rice straw and restaurant waste employing black soldier fly assisted by microbes. *Energy, 47,* 225–229. https://doi.org/10.1016/j.energy.2012.09.006.

Zimbardi, F., & Cerone, N. (2007). Energetic and environmental assessment of ethanol production from sugar and starch crops. 15Th Ebce.

Cumulative Energy Demand of Hydrogen Energy Systems

Antonio Valente, Diego Iribarren and Javier Dufour

Abstract Hydrogen energy systems are expected to play a significant role in achieving a sustainable energy sector. This requires that sustainable hydrogen options are actually available and implemented. In order to check the suitability of hydrogen under sustainability aspects, the life cycle assessment methodology is often used. In particular, global warming (i.e., carbon footprint) and cumulative energy demand (CED or energy footprint) are among the most common life-cycle indicators evaluated for hydrogen energy systems. This chapter provides a complete library of consistent (i.e., harmonised) CED values for a high number of hydrogen production options belonging to different technological categories (thermochemical, electrochemical, and biological). Overall, 71 case studies of renewable hydrogen are benchmarked—in terms of CED—against the reference case of conventional (fossil-based) hydrogen from steam reforming of natural gas. Furthermore, a correlation equation between CED and carbon footprint is calculated and applied for the estimation of harmonised CED values. The use of harmonised values allows sound comparisons by mitigating the risk of misinterpretation. The results show that electrochemical hydrogen generally performs better than thermochemical hydrogen, while biological systems show a high dispersion of values. Especially, the use of wind power as the driving energy for electrochemical hydrogen production tends to be associated with a favourable performance.

A. Valente · D. Iribarren (✉) · J. Dufour
Systems Analysis Unit, IMDEA Energy, 28935 Móstoles, Spain
e-mail: diego.iribarren@imdea.org

A. Valente · J. Dufour
Chemical and Environmental Engineering Group, Rey Juan Carlos University, 28933 Móstoles, Spain

S. S. Muthu (ed.), *Energy Footprints of the Energy Sector*,
Environmental Footprints and Eco-design of Products and Processes,
https://doi.org/10.1007/978-981-13-2457-4_2

1 Introduction

To date, fossil fuels dominate the energy outlook. However, the growing energy demand and the increasing socio-environmental concerns make the decarbonisation of energy sources a priority. Hence, within this context of unsustainability of the energy sector, low-carbon energy options are required to meet the global energy demand.

In particular, the production of energy from fossil or nuclear fuels presents significant concerns from the environmental standpoint. While fossil fuels represent one of the first causes of global warming (due to greenhouse gas emissions) and acidification (due to the release of sulphur dioxide), nuclear fuels raise concerns about safety and contamination of soil and groundwater due to radioactive waste management.

On the other hand, renewable energy has great potential all over the world to satisfy the needs of industrial, residential and transport sectors. Many benefits can be gained from the use of renewable sources under environmental, economic and health aspects. Since these energy sources are renewed through cycles much shorter than those of non-renewable fuels, they are considered inexhaustible. Their use can significantly contribute to national energy security by reducing dependence on foreign energy suppliers. Furthermore, it promotes economic development with positive consequences on employment.

Nevertheless, there are disadvantages associated with renewable energy technologies, such as a lower efficiency when compared to conventional technologies. Intermittency is another relevant issue, which potentially affects their continuous availability both throughout the day and along seasons. Given the desired stability in energy availability, the storage of surplus renewable energy (i.e., exceeding the demand) is a crucial issue widely discussed by the scientific community.

Many energy carriers are proposed in the literature, each one with its advantages and drawbacks. Among the most promising ones, hydrogen has been gaining growing attention thanks to its high potential for use in mobility or stationary applications. Nonetheless, hydrogen is still mainly used for non-energy purposes such as ammonia production or use in metallurgy (Dincer 2012).

Key advantages of hydrogen are:

- high energy content per mass unit;
- water as the only product of its oxidation;
- production through many technological pathways;
- production can take place without geographical limitations;
- it can be produced from different energy sources.

In contrast, several issues have to be taken into account for hydrogen technologies, in particular:

- low energy content per volume unit;
- lack of an adequate infrastructure for distribution;
- difficulties in hydrogen storage;

- social acceptance (being a new paradigm);
- relatively high production costs;
- as it is not directly available in pure form, hydrogen needs to be separated from its carrier (water, biomass, hydrocarbons) through technological pathways with a high energy demand.

Many of these strengths and weaknesses refer to different stages of the life cycle of hydrogen (raw material production, hydrogen production, storage, distribution, and use). Hence, when checking the techno-environmental performance of hydrogen energy systems, a life-cycle perspective needs to be followed. In this sense, Life Cycle Assessment (LCA) is a standardised methodology widely used to comprehensively evaluate the environmental performance of product systems (International Organization for Standardization 2006a, b). Through LCA, it is possible to evaluate a wide range of environmental issues, e.g. carbon footprint, energy footprint, acidification, etc. LCA is broadly applied to evaluate hydrogen energy systems (Valente et al. 2017a).

When a life-cycle perspective is used in the energy footprint assessment, a significant non-renewable energy demand could be found even for systems declared renewable. In this respect, the life-cycle energy footprint (or cumulative energy demand, CED) is seen as a useful screening indicator not only to understand the technical performance of a system but also to crosscheck its environmental performance. In fact, it is commonly linked to other environmental indicators such as the global warming impact potential (GWP or carbon footprint), especially in energy systems (Huijbregts et al. 2006).

For these reasons, this chapter focuses on the evaluation of the cumulative non-renewable energy demand (CED_{nr}) of hydrogen produced through different pathways. Emphasis is laid on renewable hydrogen energy systems and their benchmarking against conventional (fossil-based) hydrogen in terms of energy footprint. Given the comparative nature of the benchmarking study, special attention is paid to the mitigation of the misinterpretation risk associated with inconsistent methodological choices in comparative LCA (Valente et al. 2017a). Therefore, this chapter takes into account current LCA harmonisation initiatives (Valente et al. 2017b, 2018a, b) in order to provide consistent energy footprints of alternative hydrogen options. Figure 1 shows the roadmap of the chapter.

2 Materials and Methods

The goal of this chapter is to robustly frame the range of hydrogen energy options by providing the to-date most complete library of harmonised life-cycle energy footprints of hydrogen. With this aim, a large set of LCA case studies of hydrogen is needed. As shown in Fig. 2, the starting point is the extensive review of LCA studies of hydrogen energy systems performed by Valente et al. (2017a). Then a screening of the involved case studies is performed, followed by the definition of a

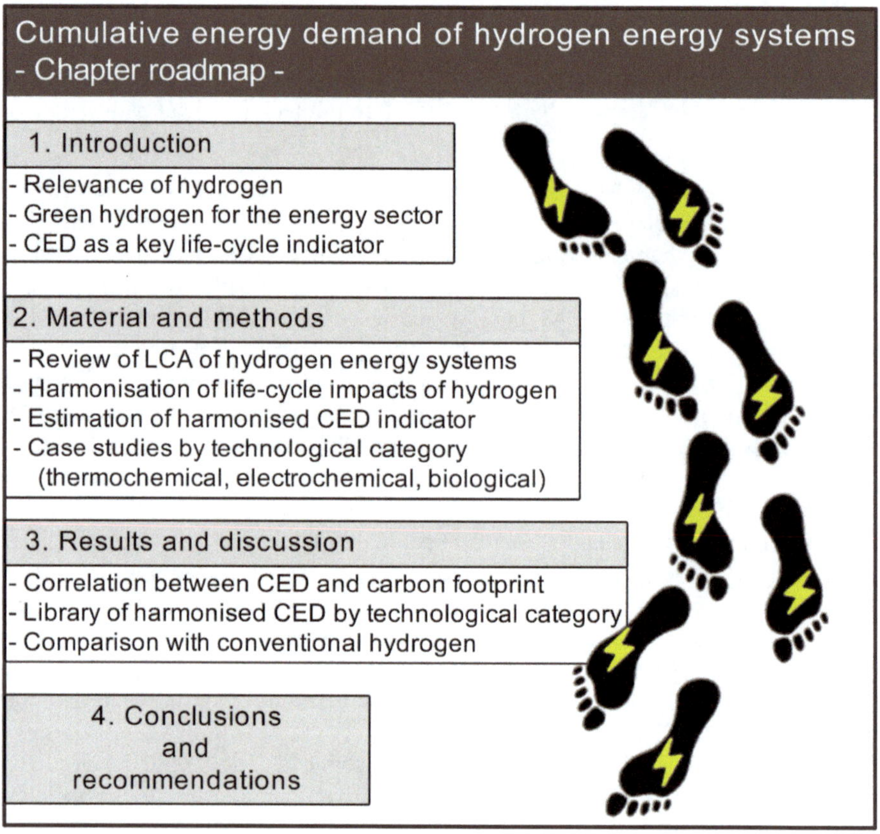

Fig. 1 Structure of the chapter

harmonisation framework dealing with the methodological choices behind the evaluation of carbon footprint (GWP) and energy footprint (CED_{nr}). Provided that a strong correlation is found between these two indicators, one indicator could be estimated from the other by simply applying a correlation equation. The final outcome of the work is an extensive library of robust energy footprints of hydrogen.

2.1 Review of LCA Studies

As defined in the standards (International Organization for Standardization 2006a, b), the LCA methodology involves four main stages. In the first one—goal and scope definition–, key aspects such as the objectives of the study, its restrictions and assumptions, the functional unit and system boundaries are addressed.

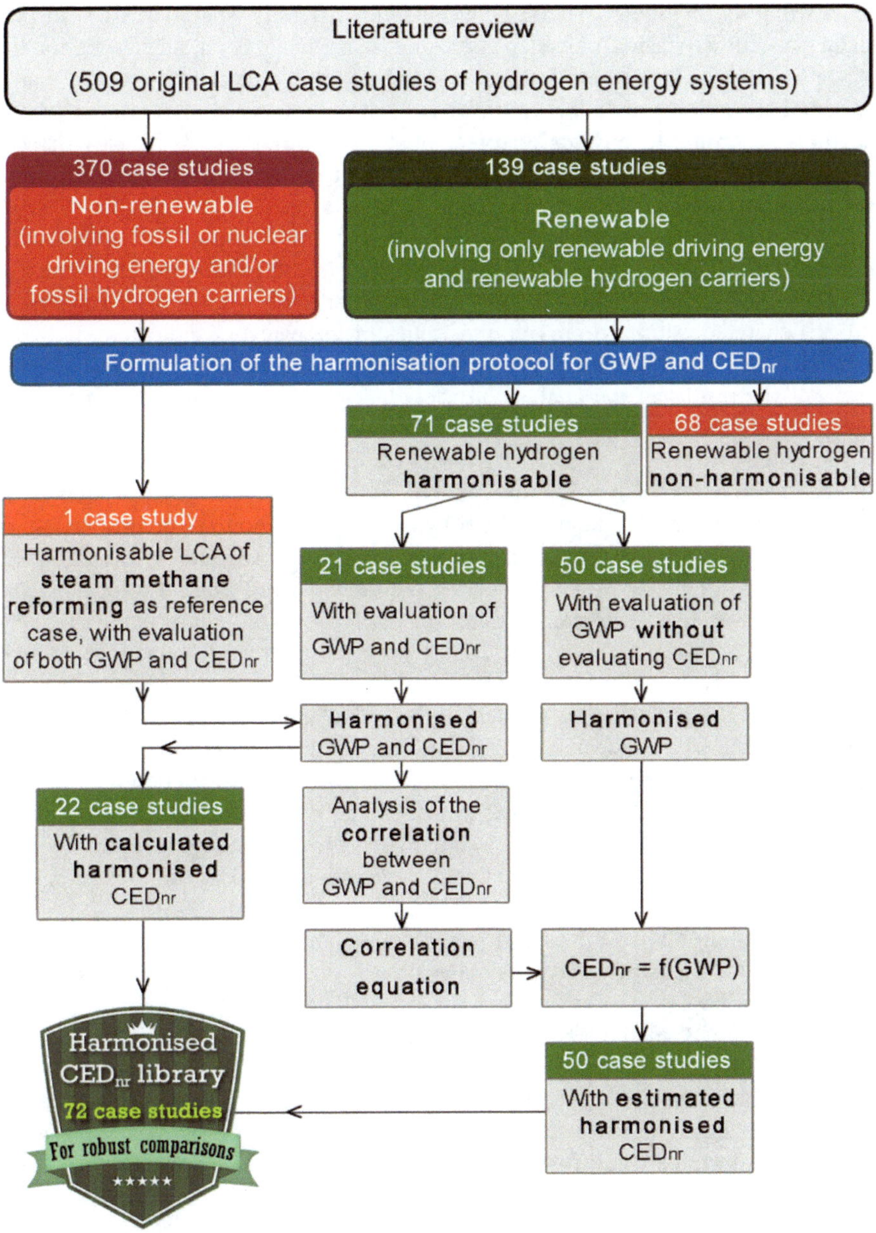

Fig. 2 Procedure followed to build the library of harmonised energy footprints

Life cycle inventory analysis is the second stage. In this stage, input and output flows of materials and energy are quantified for the system under study. Furthermore, for those systems presenting more than one product or function, the distribution of inventory data between functions has to be addressed. In particular, system expansion considers the additional functions as substitutes for the conventional ones, whose environmental burdens are avoided. On the other hand, the use of allocation approaches involves the distribution of inventory data between functions according to physical or other relationships (e.g., mass, energy or economic allocation). In this regard, LCA standards prioritise process subdivision and system expansion over the use of allocation approaches.

Three mandatory phases are involved in the third step (life cycle impact assessment, LCIA): (i) selection of impact categories, indicators and characterisation models; (ii) classification to associate inventory data with impact categories; and (iii) characterisation to provide the values of the category indicators.

Finally, the fourth stage of LCA is results interpretation, which summarises and discusses the results of the analysis in accordance with the goal and scope of the study.

The extensive review performed by Valente et al. (2017a) focuses on methodological choices in LCA of hydrogen energy systems. The review reports an increasing interest in this field for both hydrogen production and hydrogen use in mobility and stationary applications. A sample of 509 case studies was analysed, finding that most of the authors consider thermochemical (mainly reforming and gasification) or electrochemical (water electrolysis) hydrogen-production technologies. In contrast, few authors address biological hydrogen production methods (fermentation and bio-photolysis).

Regarding thermochemical studies, since steam methane reforming (SMR) is commonly used as the reference technology for comparative purposes, natural gas is the most common feedstock considered (conventional hydrogen donor), ahead of biomass. Concerning electrochemical hydrogen, the most typical electricity sources considered for water electrolysis are wind power, grid electricity, and photovoltaic power. Finally, for the biological category, microalgae arise as the most common hydrogen donor considered.

Overall, about 30% of the case studies involve renewable hydrogen production, i.e. a renewable feedstock and renewable power sources (driving energy) for the conversion process.

Regarding LCA methodological aspects, 45% of the case studies include the hydrogen use stage in the system's boundaries (cradle-to-grave approach), while 20% of the case studies consider hydrogen storage (mainly compressed) as the final stage.

Around 20% of the reviewed case studies deal with systems presenting more than one product, i.e. multifunctional systems. In this regard, electricity, animal fodder and heat are the main products involved in addition to hydrogen. System expansion is the main methodological approach followed when facing multifunctionality, thus discounting the avoided burdens associated with the co-products.

As regards environmental impact categories, Fig. 3 shows those most commonly evaluated in LCA studies of hydrogen energy systems. Out of more than 50 impact categories found in the reviewed literature, the four categories included in Fig. 3 (viz., carbon footprint, energy footprint, acidification, and eutrophication) are the only ones with a number of occurrences above 30.

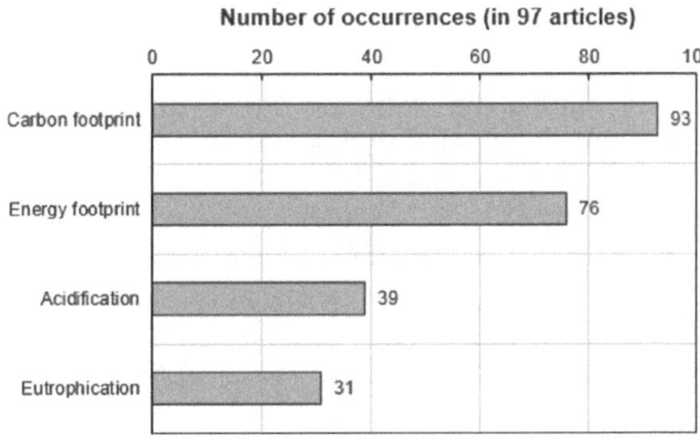

Fig. 3 Main life-cycle impact categories evaluated for hydrogen energy systems

Since global warming represents a key concern when evaluating the environmental performance of a system, GWP (i.e., carbon footprint) clearly arises as the most common impact category evaluated, being included in almost all the reviewed studies. Another highly relevant impact category is the cumulative energy demand (CED, typically expressed in MJ), which is understood as the energy footprint since it quantifies the total primary energy input of a product system. This energy footprint can refer to the total CED or to sub-components such as fossil CED, nuclear CED, non-renewable CED, renewable CED, etc. Regarding LCIA methods, IPCC (Myhre et al. 2013) and VDI (2012) are the methods typically used to evaluate GWP and CED, respectively. Other relevant categories such as acidification are evaluated mainly through the CML method (Guinée et al. 2001).

Within this context, when comparing the life-cycle results reported in different studies for alternative hydrogen options, it is crucial to pay attention to the methodological choices made in each analysis. In other words, there is a need for consistent choices when comparing LCA case studies. This need is the core of the harmonisation initiative for LCA of hydrogen (Valente et al. 2017b, 2018a, 2018b).

2.2 *Overview of Harmonisation Initiatives*

Even though LCA is a standardised methodology, practitioners are relatively free to make their own choices on different methodological aspects such as functional unit, system boundaries, multifunctionality approach, LCIA method, etc. On the one hand, this flexibility represents a strength of the methodology by facilitating its adaptation to different situations of e.g. data availability. On the other hand, it gives rise to misinterpretation risk due to concerns about the actual comparability of case studies even when evaluating the same product. In this regard, when the goal of the analysis is a comparison of different options, the methodological choices made to evaluate the impacts of the systems under study need to be homogeneous. In this situation, in order to mitigate misinterpretation, the harmonisation of methodological choices is essential.

The FC-HyGuide (Lozanovski et al. 2011) offers specific guidelines for LCA practitioners willing to evaluate hydrogen production systems. This guide provides relevant recommendations e.g. regarding the specification of final hydrogen conditions (pressure, temperature, and purity). In this respect, a key requirement for comparative LCA studies is that the comparison between different hydrogen production systems shall be carried out considering the same hydrogen conditions.

Beyond (and in agreement with) general ISO recommendations (International Organization for Standardization 2006a, b), specific recommendations provided by the FC-HyGuide document (Lozanovski et al. 2011) and the main trends observed in the specific LCA literature (Valente et al. 2017a), there are hydrogen-specific protocols to harmonise life-cycle indicators. To date, these protocols deal with GWP (Valente et al. 2017b), CED_{nr} (Valente et al. 2018a), and acidification (2018b). They have already been applied to conventional hydrogen from SMR as well as to a wide range of renewable hydrogen options found in the scientific literature, making available a library of harmonised indicators for robust benchmarking and comparative studies.

The LCA harmonisation protocols for hydrogen are based on consistent methodological choices for a common hydrogen production framework (Fig. 4). Each protocol provides specific instructions for its step-by-step application, distinguishing four main blocks. The first block defines the LCIA method, the general modelling approach, and (partly) system boundaries. The second block sets the functional unit, while the third one deals with the multifunctionality approach. Finally, the fourth block defines the final pressure of hydrogen and completes the system's boundaries by adding the compression stage and capital goods (when needed).

Except for the conditioning stage, for which a default hydrogen compression technique is considered, the available protocols rely on the original technical

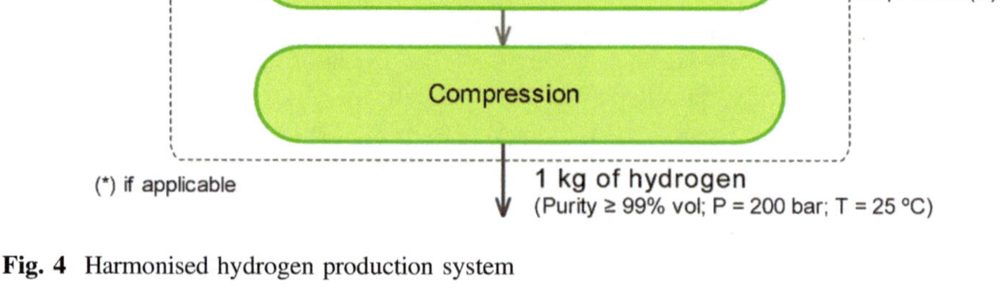

Fig. 4 Harmonised hydrogen production system

choices found in the specific case study under harmonisation. Therefore, the harmonisation exercise tends to preserve the technological features defined by the original authors. For instance, when calculating the harmonised impacts, the same type of electricity originally considered is consistently applied to add the impacts from compression, as well as to discount the burdens avoided from the electricity co-produced (when needed). Non-methodological features such as capacity, feedstock, geographical scope, etc. are thus retained.

When performing case-by-case comparisons of the carbon/energy/acidification footprint of hydrogen, the use of harmonised life-cycle indicators has already proven to mitigate misinterpretation concerns (Valente et al. 2018b). Hence, this chapter relies on the use of harmonised values in order to appropriately benchmark hydrogen options in terms of energy footprint.

2.3 Estimation of Harmonised CED_{nr}

When focusing on GWP and CED_{nr} of energy systems, a high correlation is expected between these two life-cycle indicators (Huijbregts et al. 2006). In other words, it can be claimed that the non-renewable energy footprint of an energy product is largely responsible for its carbon footprint. In the specific case of hydrogen, this relationship has been verified by Valente et al. (2018a) using harmonised values.

The actual correlation between two life-cycle indicators would enable analysts to estimate one indicator from the other by simply applying a correlation equation. In fact, this idea has recently been explored using acidification and GWP as life-cycle indicators (Valente et al. 2018b). Despite the correlation generally expected between these indicators for energy production processes (Huijbregts et al. 2006), Valente et al. (2018b) found that the idea of estimating acidification from GWP (or vice versa) should be rejected when dealing with (renewable) hydrogen energy systems due to the lack of correlation.

However, the situation is completely different when focusing on the correlation between GWP and CED_{nr}. In this regard, a strong correlation between these two indicators has already been reported in Valente et al. (2018a). Consequently, this chapter not only uses the existing library of harmonised CED_{nr} of hydrogen (Valente et al. 2018a), but it also enlarges this library by determining a GWP/CED_{nr} correlation equation and using it to estimate harmonised energy footprints from available harmonised carbon footprints.

2.4 Sample of Case Studies

This section presents the sample of case studies of hydrogen energy systems for which robust values of the CED_{nr} indicator are provided later (Sect. 3). The sample is divided into three main technological categories: thermochemical (Table 1), electrochemical (Table 2), and biological (Table 3). It should be noted that, in addition to the case studies of thermochemical renewable hydrogen, Table 1 also includes the reference case study of steam reforming of natural gas (SMR) as the conventional non-renewable hydrogen energy system. Therefore, the final number of case studies in the whole sample is 72 (71 renewable + 1 fossil). The different

Table 1 Case studies of hydrogen within the thermochemical category

Code	Original reference	Hydrogen production process
SMR	Susmozas et al. (2013)	Steam reforming of natural gas (reference case study)
TC1	Dufour et al. (2012)	$NiFe_2O_4$ thermochemical cycle (heat from solar concentrator)
TC2	Dufour et al. (2012)	ZnO thermochemical cycle (heat from solar concentrator)
BR1	Hajjaji et al. (2013)	Bioethanol reforming (from wheat grains fermentation)
BR2	Marquevich et al. (2002)	Bio-oil reforming (rape-seed oil)
BR3	Marquevich et al. (2002)	Bio-oil reforming (palm oil)
BR4	Authayanun et al. (2015)	Bioethanol (56%) + CH_4 (44%) reforming (from cassava fermentation)
BR5	Authayanun et al. (2015)	Bioethanol reforming (from cassava fermentation)
BR6	Hajjaji et al. (2013)	Autothermal reforming of bioethanol (from wheat grains fermentation)
BR7	Hajjaji et al. (2013)	Autothermal reforming of biomethane (from anaerobic digestion of cattle manure)
BR8	Wulf and Kaltschmitt (2013)	Biomethane reforming (from anaerobic digestion of non-food waste)
BR9	Wulf and Kaltschmitt (2013)	Biomethane reforming (from anaerobic digestion of German substrate mix)
BR10	Hajjaji et al. (2016)	Biogas reforming (from anaerobic digestion of farm waste)
BR11	Hajjaji et al. (2013)	Biomethane reforming (from anaerobic digestion of cattle manure)
BR12	Heracleous (2011)	Bio-oil reforming (from fast pyrolysis of wood chips)
BR13	Heracleous (2011)	Bio-oil reforming (from fast pyrolysis of willow)
BR14	Susmozas et al. (2015)	Bio-oil reforming (from fast pyrolysis of poplar)
BR15	Hajjaji et al. (2013)	Partial oxidation of biomethane (from anaerobic digestion of cattle manure)
BG1	Iribarren et al. (2014)	Biomass gasification (short-rotation poplar)
BG2	Wulf and Kaltschmitt (2013)	Biomass gasification (willow)
BG3	Wulf and Kaltschmitt (2012)	Biomass gasification (wood chips)
BG4	Susmozas et al. (2013)	Biomass gasification (poplar)

(continued)

Table 1 (continued)

Code	Original reference	Hydrogen production process
BG5	Weinberg and Kaltschmitt (2013)	Biomass gasification (woody biomass)
BG6	Simons and Bauer (2011)	Biomass gasification (woody biomass)
BG7	Martín-Gamboa et al. (2016)	Biomass gasification (grape pruning waste)
BG8	Koroneos et al. (2008)	Biomass gasification (woody biomass)
BG9	Susmozas et al. (2016)	Biomass gasification + CO_2 capture (short-rotation poplar)

Table 2 Case studies of hydrogen within the electrochemical category

Code	Original reference	Hydrogen production process
AE1	Ramos Pereira and Coelho (2013)	Alkaline electrolysis (wind power)
AE2	Granovskii et al. (2006)	Alkaline electrolysis (wind power)
AE3	Spath and Mann (2004)	Alkaline electrolysis (wind power)
AE4	Granovskii et al. (2007)	Alkaline electrolysis (wind power)
AE5	Khan et al. (2005)	Alkaline electrolysis (wind power)
AE6	Miotti et al. (2017)	Alkaline electrolysis (wind power)
AE7	Cetinkaya et al. (2012)	Alkaline electrolysis (wind power)
AE8	Simons and Bauer (2011)	Alkaline electrolysis (wind power)
AE9	Koj et al. (2015)	Alkaline electrolysis (asbestos membrane; wind power)
AE10	Koj et al. (2015)	Alkaline electrolysis (advanced membrane; wind power)
AE11	Koj et al. (2015)	Alkaline electrolysis (advanced membrane; wind power)
AE12	Suleman et al. (2015)	Alkaline electrolysis (NaCl cell; wind power)
AE13	Lee et al. (2010)	Alkaline electrolysis (wind power)
AE14	Biswas et al. (2013)	Alkaline electrolysis (wind power)
AE15	Koroneos et al. (2004)	Alkaline electrolysis (wind power)
AE16	Simons and Bauer (2011)	Alkaline electrolysis (PV power)
AE17	Cetinkaya et al. (2012)	Alkaline electrolysis (PV power)
AE18	Ramos Pereira and Coelho (2013)	Alkaline electrolysis (PV power)
AE19	Granovskii et al. (2006)	Alkaline electrolysis (PV power)

(continued)

Table 2 (continued)

Code	Original reference	Hydrogen production process
AE20	Granovskii et al. (2007)	Alkaline electrolysis (PV power)
AE21	Suleman et al. (2015)	Alkaline electrolysis (NaCl cell; PV power)
AE22	Koroneos et al. (2004)	Alkaline electrolysis (PV power)
AE23	Lombardi et al. (2011)	Alkaline electrolysis (PV power)
AE24	Koroneos et al. (2004)	Alkaline electrolysis (concentrated solar power)
AE25	Simons and Bauer (2011)	Alkaline electrolysis (concentrated solar power)
AE26	Simons and Bauer (2011)	Alkaline electrolysis (hydropower)
AE27	Valente et al. (2015)	Alkaline electrolysis (hydropower)
AE28	Koroneos et al. (2004)	Alkaline electrolysis (hydropower)
AE29	Lombardi et al. (2011)	Alkaline electrolysis (mini-hydropower)
AE30	Koroneos et al. (2004)	Alkaline electrolysis (biomass gasification power)
AE31	Mori et al. (2014)	Alkaline electrolysis (renewable power)
AE32	Wulf and Kaltschmitt (2012)	Alkaline electrolysis (renewable power)
PE1	Reiter and Lindorfer (2015)	PEM electrolysis (wind power)
PE2	Reiter and Lindorfer (2015)	PEM electrolysis (PV power)
HE1	Patyk et al. (2013)	High-temperature alkaline electrolysis (wind power)
HE2	Patyk et al. (2013)	High-temperature alkaline electrolysis (intermittent wind power)
HE3	Patyk et al. (2013)	High-temperature alkaline electrolysis (intermittent wind power with biogas reforming back-up)

Table 3 Case studies of hydrogen within the biological category

Code	Reference	Hydrogen production process
DF1	Manish and Banerjee (2008)	Dark fermentation (sugarcane)
DF2	Pacheco et al. (2015)	Dark fermentation (microalgae)
DF3	Pacheco et al. (2015)	Dark fermentation (microalgae)
PF1	Manish and Banerjee (2008)	Photo-fermentation (sugarcane)
TF1	Djomo and Blumberga (2011)	Two-stage fermentation (wheat straw)
TF2	Djomo and Blumberga (2011)	Two-stage fermentation (potato peels)
TF3	Djomo and Blumberga (2011)	Two-stage fermentation (sweet sorghum stalk)
TF4	Manish and Banerjee (2008)	Two-stage fermentation (sugarcane)

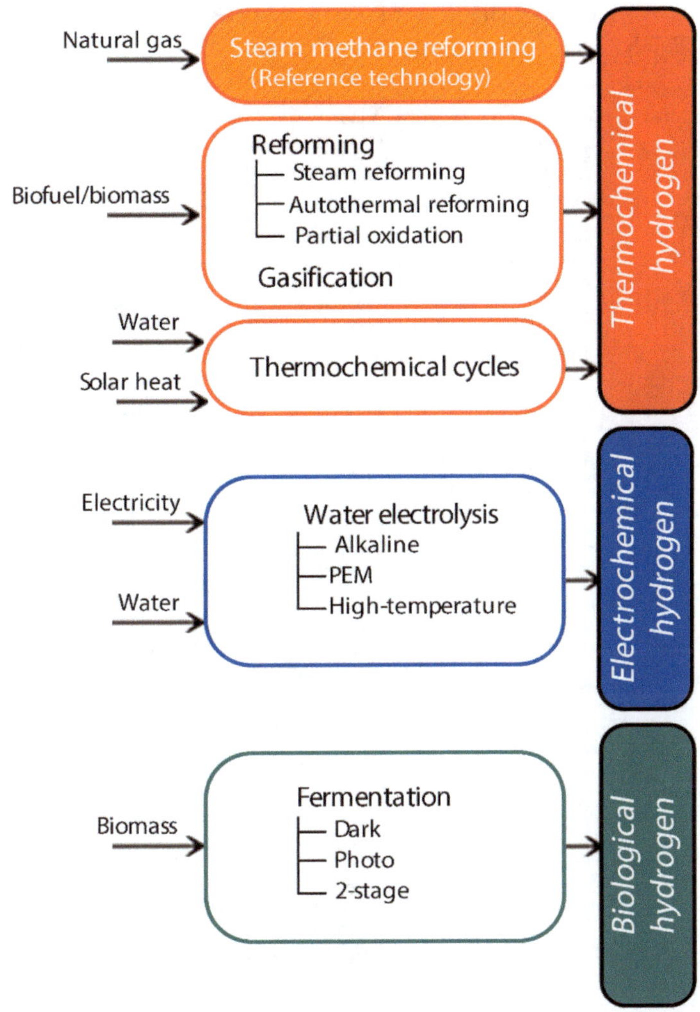

Fig. 5 Hydrogen production technologies involved in the study

hydrogen production technologies involved in the study are summarised in Fig. 5 and further described in Tables 4, 5 and 6.

Within the thermochemical category (26 renewable case studies; Table 1), biofuel reforming (BR1–BR15) is found to be the most common technology involved, with bio-oil, biomethane and bioethanol as the most typical biofuels. Furthermore, a significant number of case studies address biomass gasification (BG1–BG9), while only two case studies tackle thermochemical cycles driven by renewable energy (TC1 and TC2). The main feedstock considered is second-generation biomass based on lignocellulosic or waste matter. Fewer studies

Table 4 Description of the hydrogen production technologies involved in the study: thermochemical hydrogen

Technology	Description
Steam reforming	Endothermic process. Water in the form of steam is used to oxidise the carbon contained in the feedstock (liquid or gaseous fuel)
Partial oxidation	Exothermic process. Oxygen below the stoichiometric ratio is used to oxidise the carbon contained in the feedstock (liquid or gaseous fuel)
Autothermal reforming	Combination of steam reforming and partial oxidation technologies: the partial oxidation of the feedstock provides the heat demanded by the steam reforming process
Gasification	Conversion of a carbonaceous feedstock into syngas at elevated temperature in a gasification medium such as air, oxygen, and/or steam
Thermochemical cycles	Thermal dissociation of water in a sequence of reactions comprising different chemical reactants that are recovered at the end of the cycle. The required temperature is significantly lower than for the thermal decomposition of water in a single step

Table 5 Description of the hydrogen production technologies involved in the study: electrochemical hydrogen

Technology	Description
Alkaline water electrolysis	Dissociation of water into ions in an alkaline electrolytic solution at temperature below 80 °C (low-temperature electrolysis) under the effect of a direct current supplied to the electrodes
PEM water electrolysis	Dissociation of water into ions at temperature below 80 °C (low-temperature electrolysis) under the effect of a direct current supplied to the electrodes. The anode and the cathode are separated by a polymeric membrane as the electrolyte (acid), which allows the cationic exchange between the electrodes while preventing electron exchange
High-temperature electrolysis	Temperatures within the range 600–1000 °C are needed (e.g., solid oxide electrolysis and molten carbonate electrolysis). Reduced electricity consumption by using heat in order to meet the energy demand of the process

Table 6 Description of the hydrogen production technologies involved in the study: biological hydrogen

Technology	Description
Dark fermentation	Anaerobic bacteria are used to produce hydrogen and organic acids from glucose in dark conditions
Photo-fermentation	Photosynthetic bacteria are involved, which convert organic acids into hydrogen and CO_2 under the catalytic action of nitrogenase with anaerobic conditions, neutral pH, and mesophilic temperature range
Two-stage fermentation	Dark- and photo-fermentation are combined in sequence in order to reach higher yields of hydrogen

consider first-generation biomass or water (the latter is used as the hydrogen carrier in thermochemical cycles).

The electrochemical category involves 37 case studies of renewable hydrogen (Table 2). Alkaline water electrolysis clearly dominates this technological category (cases AE1–AE32 and three case studies of high-temperature alkaline electrolysis, HE1–HE3), while only two case studies consider proton exchange membrane (PEM) water electrolysis (PE1 and PE2). The most common source of electricity is found to be wind power, and a significant number of case studies consider photovoltaic (PV) electricity. Other types of electricity considered are hydropower, concentrated solar power, and electricity from biomass.

With regard to biological hydrogen (Table 3), a low number of case studies is found in the literature. Hence, only eight case studies are considered in this chapter. All of them involve fermentative processes, which can be further classified into dark-fermentation (DF1–DF3), photo-fermentation (PF1), and two-stage fermentation (TF1–TF4).

2.5 Contextualisation within IEA HIA Task 36

The harmonisation of life-cycle indicators of hydrogen energy systems is framed within Task 36 of the International Energy Agency (IEA) Hydrogen Implementing Agreement (HIA): "Life Cycle Sustainability Assessment of Hydrogen Energy Systems". During the period 2015–2017, the goal of this task was to facilitate decision-making in the hydrogen energy sector by providing a robust and comprehensive methodological framework for the sustainability assessment of hydrogen energy systems (Dufour et al. 2015). Experts from five different countries (Spain, Germany, Japan, Norway, and Italy) worked within the framework of IEA HIA Task 36, which involved four sub-tasks as detailed in Fig. 6.

Overall, addressing the challenge of developing a consistent life cycle sustainability assessment framework for hydrogen energy systems involves clear opportunities to enhance decision-making processes at the level of both industry and policy-makers. In addition to enable a thorough identification of specific sustainability hotspots, such a methodological framework could help hydrogen economy actors anticipate, check and prove the suitability of their hydrogen energy solutions according to the criteria required by current and future policies.

3 Results

Overall, this section provides a broad collection of robust (i.e., harmonised) CED_{nr} values for the case studies presented in Tables 1, 2 and 3.

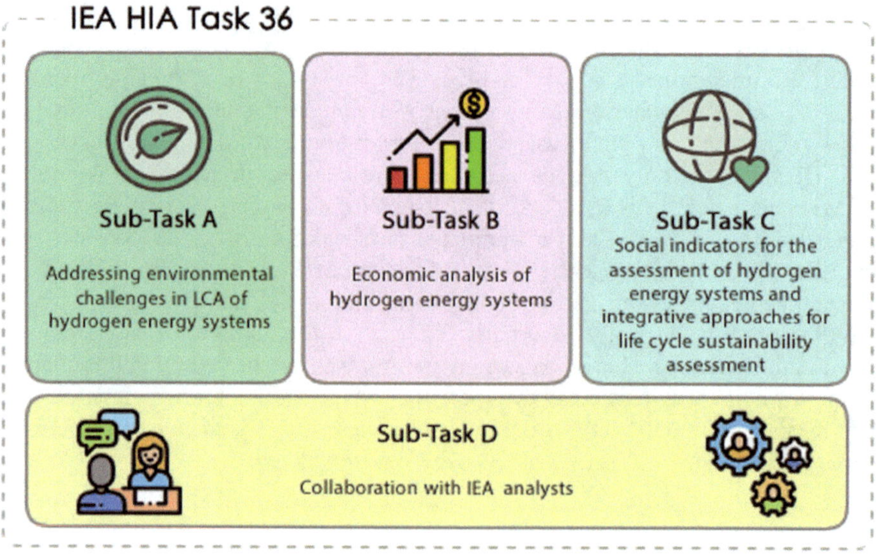

Fig. 6 IEA HIA Task 36 structure

3.1 Correlation Equation for the Estimation of CED_{nr}

From the whole sample of case studies, 50 of them do not report the evaluation of CED_{nr}, and therefore the corresponding harmonised values cannot be calculated through the step-by-step application of the protocol. For these case studies, as explained in Sect. 2.3, it is necessary to estimate their harmonised energy footprint from their harmonised carbon footprint (which is available for all the case studies within the sample) using a correlation equation between CED_{nr} and GWP. This section presents the definition of such a correlation equation.

The harmonised CED_{nr} values provided in Valente et al. (2018a) and the corresponding harmonised carbon footprints reported in Valente et al. (2017b) are used herein to determine the equation that links GWP and CED_{nr}. In this respect, the linear regression analysis in Fig. 7 conveniently shows a high correlation between GWP and CED_{nr} ($R^2 = 0.92$). Hence, the correlation equation shown in Fig. 7 can be used to estimate the harmonised CED_{nr} values in those case studies in which non-renewable energy footprints—unlike carbon footprints—are not reported.

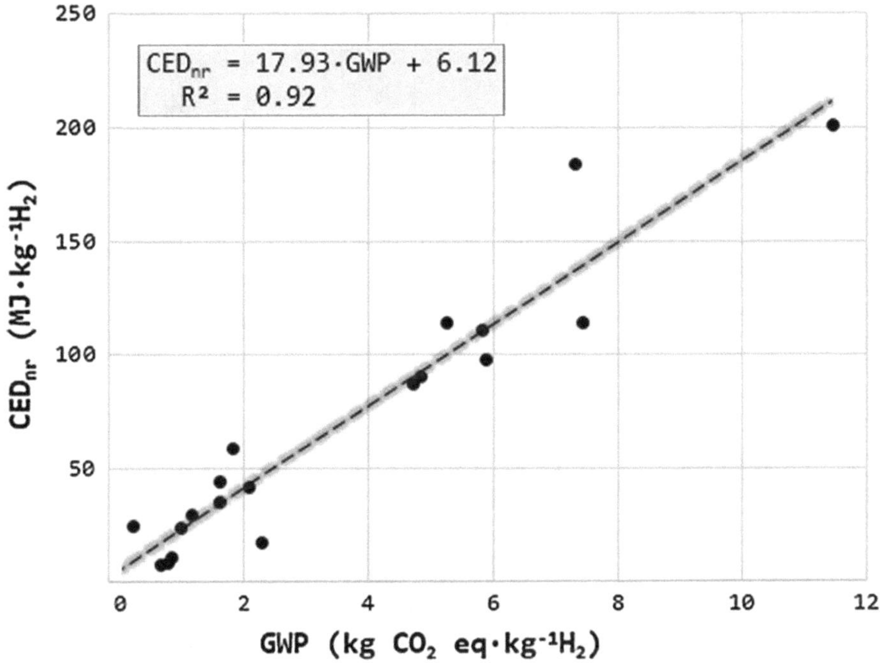

Fig. 7 Correlation between the carbon footprint and the non-renewable energy footprint of hydrogen

3.2 Library of Energy Footprints and Robust Comparison with Conventional Hydrogen

In this section, the harmonised energy footprints of a wide range of hydrogen options are reported by technological category. The use of these values rather than the original (i.e., non-harmonised) ones is considered to lead to more robust comparisons by mitigating the risk of misinterpretation.

3.2.1 Harmonised Energy Footprint of Thermochemical Hydrogen

Within the thermochemical category, 26 case studies of renewable hydrogen are included, as well as that of conventional, fossil-based hydrogen from SMR (reference case study). Table 7 reports the corresponding library of harmonised non-renewable energy footprints, distinguishing calculated and estimated CED_{nr} values. While 9 of the 27 case studies present CED_{nr} values coming from the direct application of the harmonisation protocol (Valente et al. 2018a), the remaining 18 harmonised values are estimated through the application of the correlation equation determined in Sect. 3.1.

Table 7 Library of harmonised CED_{nr} of thermochemical hydrogen

Case study code	CED^{a}_{nr} (MJ•kg^{-1} H_2)	CED^{b}_{nr} (MJ•kg^{-1} H_2)
SMR	200.9	–
TC1	–	127.7
TC2	–	125.6
BR1	–	191.7
BR2	–	137.7
BR3	–	100.0
BR4	–	93.1
BR5	–	214.0
BR6	–	184.1
BR7	–	109.6
BR8	–	130.9
BR9	–	135.2
BR10	98.2	–
BR11	–	109.9
BR12	111.2	–
BR13	114.0	–
BR14	114.7	–
BR15	–	111.4
BG1	41.9	–
BG2	–	84.8
BG3	–	63.0
BG4	25.4	–
BG5	–	80.7
BG6	–	241.9
BG7	3.0	–
BG8	–	195.9
BG9	−43.9	–

[a]According to Valente et al. (2018a)
[b]Estimated through the GWP/CED_{nr} correlation equation

The harmonised values of thermochemical hydrogen range from −44 MJ (BG9) to 242 MJ (BG6). BG9 addresses hydrogen production from poplar gasification involving a CO_2 capture system (CO_2 as an avoided product), while BG6 involves gasification of dedicated energy crops without any co-product. According to Table 7, the average CED_{nr} value for the renewable thermochemical category is 112 MJ, i.e. about 56% of the non-renewable energy footprint of conventional hydrogen from SMR.

Figure 8 shows the CED_{nr} results for renewable thermochemical hydrogen in relative terms with respect to the CED_{nr} value of the reference case (SMR). In this sense, all those case studies with a value lower than 1 perform better than SMR in terms of non-renewable energy footprint. With the aim of facilitating the

Fig. 8 Comparison between renewable thermochemical hydrogen and SMR-H_2

interpretation of the comparative study, the case studies are arranged by ascending order of energy footprint.

The results shown in Fig. 8 suggest that most of the renewable hydrogen options included in the library of thermochemical hydrogen are potential candidates to substitute hydrogen produced through the conventional technology (SMR). Hydrogen options based on biomass gasification, especially when involving more than one product, tend to show the best performance. In this regard, most of the case studies based on biomass gasification allow a CED_{nr} saving above 60% with respect to SMR. However, when this technology involves dedicated crops without any co-product, a significantly unfavourable performance is found in terms of non-renewable energy footprint.

The two case studies involving thermochemical cycle processes (TC1 and TC2) perform about 40% better than SMR. Nevertheless, both systems are assessed in the same study and the number of harmonised case studies for this type of process is too low to draw general conclusions on this technology.

Biofuel reforming shows a relatively narrow range of harmonised values. In fact, 80% of the biofuel reforming cases are within the range 30–45% of CED_{nr} savings with respect to SMR. This indicates that the biofuel reforming technology generally shows a good profile, suitable to substitute the conventional hydrogen production technology. Nevertheless, in light of the results, gasification could be preferred, especially when residual biomass is available.

3.2.2 Harmonised Energy Footprint of Electrochemical Hydrogen

The library of harmonised CED_{nr} of electrochemical hydrogen involves 37 renewable case studies (Table 8). Harmonised values are directly available in Valente et al. (2018a) for only 9 case studies, while the remaining 28 cases are harmonised by applying the correlation equation.

Table 8 Library of harmonised CED_{nr} of electrochemical hydrogen

Case study code	CED^{a}_{nr} ($MJ \cdot kg^{-1}$ H_2)	CED^{b}_{nr} ($MJ \cdot kg^{-1}$ H_2)
AE1	–	26.3
AE2	–	23.5
AE3	–	23.4
AE4	–	23.3
AE5	–	15.2
AE6	–	42.3
AE7	–	26.7
AE8	29.9	–
AE9	–	19.2
AE10	–	18.3
AE11	–	18.3
AE12	–	9.1
AE13	–	21.4
AE14	–	20.1
AE15	–	21.2
AE16	59.4	–
AE17	–	52.6
AE18	–	109.9
AE19	–	77.5
AE20	–	48.6
AE21	–	18.6
AE22	–	96.4
AE23	–	140.9
AE24	–	45.6
AE25	44.3	–
AE26	23.9	–
AE27	8.7	–
AE28	–	41.8
AE29	–	212.6
AE30	35.5	–
AE31	–	115.6
AE32	–	69.2

(continued)

Table 8 (continued)

Case study code	CED_{nr}^{a} ($MJ \cdot kg^{-1}$ H_2)	CED_{nr}^{b} ($MJ \cdot kg^{-1}$ H_2)
PE1	–	19.4
PE2	–	63.7
HE1	8.1	–
HE2	11.5	–
HE3	17.6	–

[a]According to Valente et al. (2018a)
[b]Estimated through the GWP/CED_{nr} correlation equation

The harmonised values of electrochemical hydrogen range from 8 MJ (HE1) to 213 MJ (AE29). The case AE29 involves a mini-hydropower plant as the driving energy source, while HE1 involves high-temperature electrolysis using wind power for heat and electricity. Except for AE29, the case studies within the electrochemical category perform significantly better than SMR. According to Table 8, the average value of CED_{nr} for the electrochemical category is 45 MJ, which corresponds to less than a quarter of the energy footprint of conventional hydrogen from SMR.

Figure 9 shows the harmonised CED_{nr} results of renewable electrochemical hydrogen relative to the harmonised CED_{nr} of conventional SMR hydrogen. High-temperature electrolysis (HE1–HE3) shows a very favourable performance in terms of energy footprint, performing better than PEM electrolysis (PE1 and PE2), which is probably linked to the shorter lifetime of PEM stacks (Reiter and Lindorfer 2015).

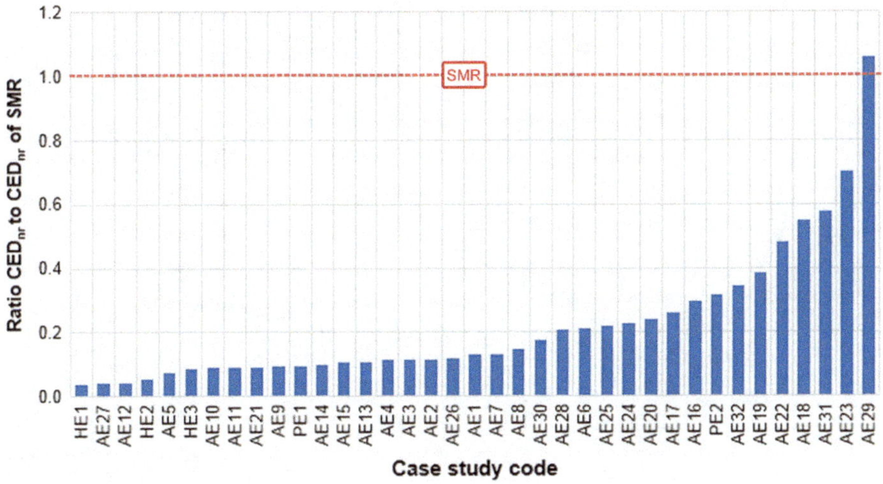

Fig. 9 Comparison between renewable electrochemical hydrogen and SMR-H_2

The main contributor to the energy footprint of electrochemical hydrogen generally is the infrastructure associated with the power source. In this respect, while the wind power-based case studies assessed generally show a CED_{nr} saving above 80% with respect to SMR, the energy footprint increases when the power source is based on solar energy. The only case study performing worse than SMR involves a small-scale hydropower plant, which suggests the relevance of capacity (i.e., scale) on the techno-environmental performance of this type of energy system.

3.2.3 Harmonised Energy Footprint of Biological Hydrogen

Table 9 reports the library of harmonised CED_{nr} of biological hydrogen, which involves 8 case studies. The library contains three values directly available in Valente et al. (2018a), while the values for the remaining five case studies are estimated by applying the correlation equation with the harmonised carbon footprints available in Valente et al. (2017b).

Although the number of case studies belonging to the biological category is significantly lower than for the other technological categories, interesting trends are also identified. In this respect, the use of microalgal biomass is associated with an unfavourable energy performance—significantly worse than SMR (see DF2 and DF3 in Fig. 10)—due to the high energy demand for microalgae growth, harvesting and drying. In contrast, two-stage fermentative processes, which involve a first stage under dark conditions (dark fermentation) and a subsequent stage of photo-fermentation, show the best performance within the biological category in terms of energy footprint.

Table 9 Library of harmonised CED_{nr} of biological hydrogen

Case study code	CED^{a}_{nr} ($MJ\cdot kg^{-1}H_2$)	CED^{b}_{nr} ($MJ\cdot kg^{-1}H_2$)
DF1	183.7	–
DF2	–	932.8
DF3	–	30,614.3
PF1	91.1	–
TF1	–	87.0
TF2	–	48.9
TF3	–	94.9
TF4	87.7	–

[a]According to Valente et al. (2018a)
[b]Estimated with the GWP/CED_{nr} correlation equation

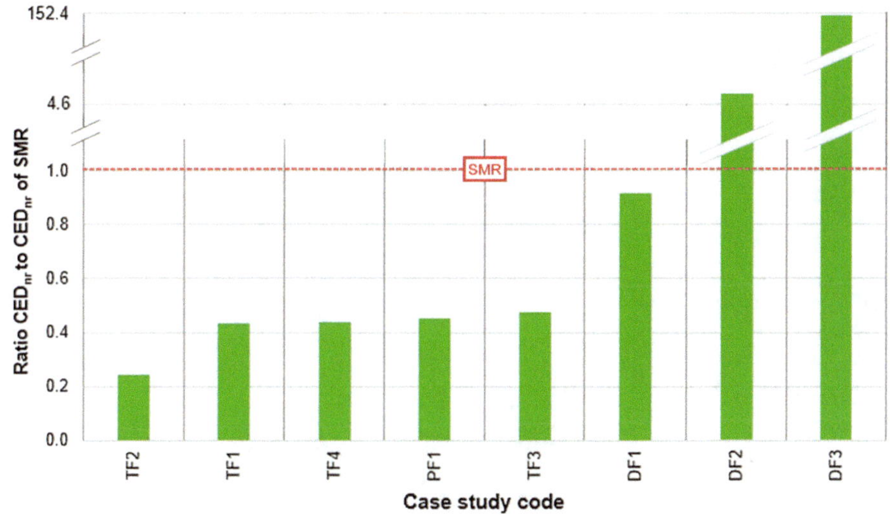

Fig. 10 Comparison between renewable biological hydrogen and SMR-H_2

3.3 Overall Picture

Thanks to the availability of a thorough library of harmonised CED$_{nr}$ of hydrogen, further results interpretation—beyond the intra-category assessment in Sect. 3.2—is possible. This is done not only at the inter-level of technological category (Fig. 11) but also at the inter-level of production technology (Fig. 12).

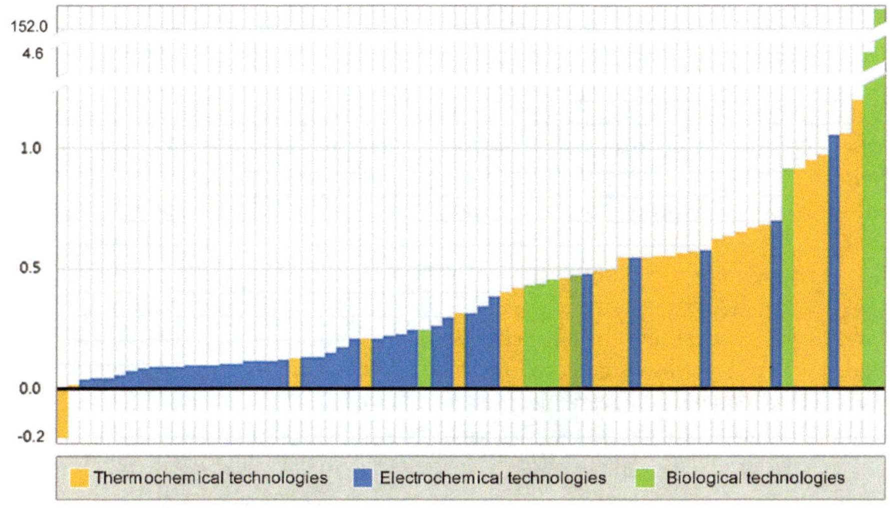

Fig. 11 Overview of energy footprint by technological category

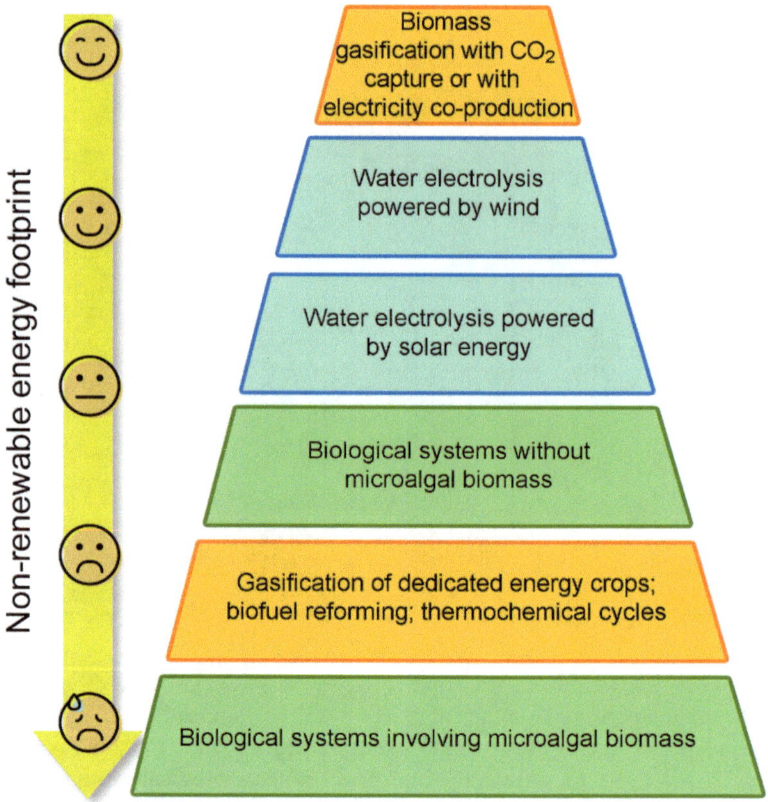

Fig. 12 Hierarchy of hydrogen production technologies based on the complete library of harmonised CED_{nr}

Figure 11 includes the whole set of 71 harmonised CED_{nr} values, still shown in relative terms with respect to the harmonised CED_{nr} of SMR and arranged in ascending order. The distribution of the values in Fig. 11 shows that electrochemical hydrogen tends to perform better than thermochemical and biological hydrogen. It is also observed that biological hydrogen shows scattered harmonised values.

It is remarkable that more than 85% of the electrochemical case studies (32 out of 37) lead to a CED_{nr} saving above 60% with respect to SMR. In fact, this saving is above 80% for ca. 65% of the electrochemical cases (24 out of 37). On the other hand, most of the case studies of thermochemical hydrogen are associated with lower energy savings. Nevertheless, more than 60% of the thermochemical case studies (16 out of 26) present CED_{nr} savings within the range 30–60%.

The average CED_{nr} saving for renewable electrochemical hydrogen is 78%, while this saving decreases to 44% for renewable thermochemical hydrogen. In contrast, the average value for the biological category does not mean any saving with respect to conventional hydrogen due to the high dispersion of the energy footprints depending on the specific case study (with some case studies characterised by very unfavourable results). Moreover, the sample of biological case studies is relatively small, and thus the considerations for this technological category should be further investigated in the future.

Overall, Fig. 12 summarises the findings about the energy performance of the renewable hydrogen production technologies included in the library of harmonised CED_{nr}. On the one hand, the least favourable technique is found within the biological category when fermentative processes involve microalgae feedstock. On the other hand, thermochemical technologies are found to be highly competitive when co-producing either electricity or goods that avoid conventional energy-intensive processes. Regarding electrochemical hydrogen, even though wind-based technologies usually outperform solar-based ones, the latter are still associated with a suitable performance when compared to other technological routes.

4 Conclusions and Recommendations

A thorough library of non-renewable energy footprints of hydrogen is now available. As a key feature, the CED_{nr} values included in this library are methodologically consistent, i.e. they are reported on the basis of harmonised LCA methodological choices regarding e.g. functional unit, system boundaries, multifunctionality approach, and final hydrogen conditions. As a result, robust comparisons between different hydrogen options—including their benchmarking against conventional, fossil-based hydrogen from SMR—can be made. In fact, future LCA practitioners willing to perform robust comparisons between hydrogen energy systems are highly recommended to use the harmonised values reported in this chapter.

The use of harmonised energy footprints facilitates the identification of the potentially most suitable hydrogen production technologies when it comes to substituting conventional hydrogen production. In this sense, electrochemical hydrogen produced via water electrolysis using wind power is generally characterised by a very favourable profile in terms of energy footprint. Even though electrolysis powered by solar energy shows a relatively worse energy footprint, it is also identified as a suitable option.

Thermochemical hydrogen presents a wide range of technologies available for the production of hydrogen. Among them, biomass gasification generally shows a favourable profile, especially when involving more products besides hydrogen. Hence, this type of technology could be considered a potential candidate to substitute the conventional hydrogen technology. Other thermochemical technologies such as biofuel reforming and thermochemical cycles show less evident benefits in terms of energy footprint when benchmarked against hydrogen produced through steam reforming of natural gas.

The case studies belonging to the biological category, which involve fermentative processes, are associated with highly scattered energy footprints. The use of microalgae as the feedstock arises as an unfavourable alternative due to the high energy requirements of cultivation and harvesting.

Overall, electrochemical hydrogen shows a more favourable energy footprint than the other technological categories (i.e., thermochemical and biological hydrogen). Nevertheless, some technical features—such as the co-production of electricity or energy-intensive goods—can significantly improve the performance of thermochemical options, leading to energy footprints even better than those of electrochemical hydrogen.

Acknowledgements This research has been partly supported by the Spanish Ministry of Economy, Industry and Competitiveness (ENE2015-74607-JIN AEI/FEDER/UE) and the Regional Government of Madrid (S2013/MAE-2882 and S2013/ABI-2783). This work is framed within Task 36 of the International Energy Agency (IEA) Hydrogen Implementing Agreement (HIA).

References

Authayanun, S., Suwanmanee, U., & Arpornwichanop, A. (2015). Enhancement of dilute bio-ethanol steam reforming for a proton exchange membrane fuel cell system by using methane as co-reactant: Performance and life cycle assessment. *International Journal of Hydrogen Energy, 40,* 12144–12153.

Biswas, W. K., Thompson, B. C., & Islam, M. N. (2013). Environmental life cycle feasibility assessment of hydrogen as an automotive fuel in Western Australia. *International Journal of Hydrogen Energy, 38,* 246–254.

Cetinkaya, E., Dincer, I., & Naterer, G. F. (2012). Life cycle assessment of various hydrogen production methods. *International Journal of Hydrogen Energy, 37,* 2071–2080.

Dincer, I. (2012). Green methods for hydrogen production. *International Journal of Hydrogen Energy, 37,* 1954–1971.

Djomo, S. N., & Blumberga, D. (2011). Comparative life cycle assessment of three biohydrogen pathways. *Bioresource technology, 102,* 2684–2694.

Dufour, J., Serrano, D. P., Gálvez, J. L., González, A., Soria, E., & Fierro, J. L. G. (2012). Life cycle assessment of alternatives for hydrogen production from renewable and fossil sources. *International Journal of Hydrogen Energy, 37,* 1173–1183.

Dufour, J., Iribarren, D., Zapp, P., Ishimoto, Y., Espegren. K. A., & Masoni, P. (2015). Towards a harmonised life cycle sustainability assessment framework for hydrogen energy systems: IEA HIA Task 36. In *Organising Committee HYPOTHESIS XI, editors. Hydrogen Power Theoretical and Engineering Solutions International Symposium 2015 —Book of Abstracts* (p. 131). Móstoles: Rey Juan Carlos University.

Granovskii, M., Dincer, I., & Rosen, M. A. (2006). Life cycle assessment of hydrogen fuel cell and gasoline vehicles. *International Journal of Hydrogen Energy, 31,* 337–352.

Granovskii, M., Dincer, I., & Rosen, M. A. (2007). Exergetic life cycle assessment of hydrogen production from renewables. *Journal of Power Sources, 167,* 461–471.

Guinée, J. B., Gorrée, M., Heijungs, R., Huppes, G., Kleijn, R., de Koning, A., et al. (2001). *Life cycle assessment—An operational guide to the ISO standards.* Leiden: Centre of Environmental Science.

Hajjaji, N., Pons, M. N., Renaudin, V., & Houas, A. (2013). Comparative life cycle assessment of eight alternatives for hydrogen production from renewable and fossil feedstock. *Journal of Cleaner Production, 44,* 177–189.

Hajjaji, N., Martinez, S., Trably, E., Steyer, J. P., & Helias, A. (2016). Life cycle assessment of hydrogen production from biogas reforming. *International Journal of Hydrogen Energy, 41,* 6064–6075.

Heracleous, E. (2011). Well-to-Wheels analysis of hydrogen production from bio-oil reforming for use in internal combustion engines. *International Journal of Hydrogen Energy, 36,* 11501–11511.

Huijbregts, M. A. J., Rombouts, L. J. A., Hellweg, S., Frischknecht, R., Hendriks, A. J., van de Meent, D., et al. (2006). Is cumulative fossil energy demand a useful indicator for the environmental performance of products? *Environmental Science and Technology, 40,* 641–648.

International Organization for Standardization. (2006a). *ISO 14040:2006 Environmental management—Life cycle assessment—Principles and framework.* Geneva: ISO.

International Organization for Standardization. (2006b). *ISO 14044:2006 Environmental management—Life cycle assessment—Requirements and guidelines.* Geneva: ISO.

Iribarren, D., Susmozas, A., Petrakopoulou, F., & Dufour, J. (2014). Environmental and exergetic evaluation of hydrogen production via lignocellulosic biomass gasification. *Journal of Cleaner Production, 69,* 165–175.

Khan, F. I., Hawboldt, K., & Iqbal, M. T. (2005). Life Cycle Analysis of wind–fuel cell integrated system. *Renew. Energy, 30,* 157–177.

Koj, J. C., Schreiber, A., Zapp, P., & Marcuello, P. (2015). Life cycle assessment of improved high pressure alkaline electrolysis. *Energy Procedia, 75,* 2871–2877.

Koroneos, C., Dompros, A., Roumbas, G., & Moussiopoulos, N. (2004). Life cycle assessment of hydrogen fuel production processes. *International Journal of Hydrogen Energy, 29,* 1443–1450.

Koroneos, C., Dompros, A., & Roumbas, G. (2008). Hydrogen production via biomass gasification–A life cycle assessment approach. *Chemical Engineering and Processing, 47,* 1261–1268.

Lee, J. Y., An, S., Cha, K., & Hur, T. (2010). Life cycle environmental and economic analyses of a hydrogen station with wind energy. *International Journal of Hydrogen Energy, 35,* 2213–2225.

Lombardi, L., Carnevale, E., & Corti, A. (2011). Life cycle assessment of different hypotheses of hydrogen production for vehicle fuel cells fuelling. *International of Journal Energy Environment Engineering, 2,* 63–78.

Lozanovski, A., Schuller, O., & Faltenbacher, M. (2011). *Guidance document for performing LCA on hydrogen production systems.* Brussels: FCH JU.

Manish, S., & Banerjee, R. (2008). Comparison of biohydrogen production processes. *International Journal of Hydrogen Energy, 33,* 279–286.

Marquevich, M., Sonnemann, G. W., Castells, F., & Montané, D. (2002). Life cycle inventory analysis of hydrogen production by the steam-reforming process: comparison between vegetable oils and fossil fuels as feedstock. *Green Chemistry, 4,* 414–423.

Martín-Gamboa, M., Iribarren, D., Susmozas, A., & Dufour, J. (2016). Delving into sensible measures to enhance the environmental performance of biohydrogen: A quantitative approach based on process simulation, life cycle assessment and data envelopment analysis. *Bioresource Technology, 214,* 376–385.

Miotti, M., Hofer, J., & Bauer, C. (2017). Integrated environmental and economic assessment of current and future fuel cell vehicles. *International Journal of Life Cycle Assessment, 22,* 94–110.

Mori, M., Jensterle, M., Mržljak, T., & Drobnič, B. (2014). Life-cycle assessment of a hydrogen-based uninterruptible power supply system using renewable energy. *International Journal of Life Cycle Assessment, 19,* 1810–1822.

Myhre, G., Shindell, D., Bréon, F. M., Collins, W., Fuglestvedt, J., Huang, J., et al. (2013). Anthropogenic and natural radiative forcing. In T. F. Stocker, D. Qin, G. K. Plattner, M. Tignor, S. K. Allen, & J. Boschung (Eds.), *Climate Change 2013: The Physical Science Basis—Contribution of Working Group I to the Fifth Assessment Report of the Intergovernmental Panel on Climate Change* (pp. 659–740). Cambridge: Cambridge University Press.

Pacheco, R., Ferreira, A. F., Pinto, T., Nobre, B. P., Loureiro, D., Moura, P., et al. (2015). The production of pigments & hydrogen through a *Spirogyra* sp. biorefinery. *Energy Conversion and Management, 89,* 789–797.

Patyk, A., Bachmann, T. M., & Brisse, A. (2013). Life cycle assessment of H_2 generation with high temperature electrolysis. *International Journal of Hydrogen Energy, 38,* 3865–3880.

Ramos Pereira, S., & Coelho, M. C. (2013). Life cycle analysis of hydrogen—A well-to-wheels analysis for Portugal. *International Journal of Hydrogen Energy, 38,* 2029–2038.

Reiter, G., & Lindorfer, J. (2015). Global warming potential of hydrogen and methane production from renewable electricity via power-to-gas technology. *International Journal of Life Cycle Assessment, 20,* 477–489.

Simons, A., & Bauer, C. (2011). Life cycle assessment of hydrogen production. In A. Wokaun & E. Wilhelm (Eds.), *Transition to Hydrogen—Pathways toward Clean Transportation* (pp. 13–57). Cambridge: Cambridge University Press.

Spath, P. L., & Mann, M. K. (2004). *Life cycle assessment of renewable hydrogen production via wind/electrolysis*. Golden: US National Renewable Energy Laboratory.

Suleman, F., Dincer, I., & Agelin-Chaab, M. (2015). Environmental impact assessment and comparison of some hydrogen production options. *International Journal of Hydrogen Energy, 40,* 6976–6987.

Susmozas, A., Iribarren, D., & Dufour, J. (2013). Life-cycle performance of indirect biomass gasification as a green alternative to steam methane reforming for hydrogen production. *International Journal of Hydrogen Energy, 38,* 9961–9972.

Susmozas, A., Iribarren, D., & Dufour, J. (2015). Assessing the life-cycle performance of hydrogen production via biofuel reforming in Europe. *Resources, 4,* 398–411.

Susmozas, A., Iribarren, D., Zapp, P., Linβen, J., & Dufour, J. (2016). Life-cycle performance of hydrogen production via indirect biomass gasification with CO_2 capture. *International Journal of Hydrogen Energy, 41,* 19484–19491.

Valente, A., Iribarren, D., Dufour, J., & Spazzafumo, G. (2015). Life-cycle performance of hydrogen as an energy management solution in hydropower plants: A case study in Central Italy. *International Journal of Hydrogen Energy, 40,* 16660–16672.

Valente, A., Iribarren, D., & Dufour, J. (2017a). Life cycle assessment of hydrogen energy systems: a review of methodological choices. *International Journal of Life Cycle Assessment, 22,* 346–363.

Valente, A., Iribarren, D., & Dufour, J. (2017b). Harmonised life-cycle global warming impact of renewable hydrogen. *Journal of Cleaner Production, 149,* 762–772.

Valente, A., Iribarren, D., & Dufour, J. (2018a). Harmonising the cumulative energy demand of renewable hydrogen for robust comparative life-cycle studies. *Journal of Cleaner Production, 175,* 384–393.

Valente, A., Iribarren, D., & Dufour, J. (2018b). Harmonising methodological choices in life cycle assessment of hydrogen: a focus on acidification and renewable hydrogen. *International Journal of Hydrogen Energy* (in press). https://doi.org/10.1016/j.ijhydene.2018.03.101.

VDI. (2012). *VDI guideline 4600: Cumulative energy demand (KEA)—Terms, definitions, methods of calculation*. Düsseldorf: Verein Deutscher Ingenieure.

Weinberg, J., & Kaltschmitt, M. (2013). Life cycle assessment of mobility options using wood based fuels—comparison of selected environmental effects and costs. *Bioresource Technology, 150,* 420–428.

Wulf, C., & Kaltschmitt, M. (2012). Life cycle assessment of hydrogen supply chain with special attention on hydrogen refuelling stations. *International Journal of Hydrogen Energy, 37,* 16711–16721.

Wulf, C., & Kaltschmitt, M. (2013). Life cycle assessment of biohydrogen production as a transportation fuel in Germany. *Bioresource Technology, 150,* 466–475.

Energy Footprint of India: Scope for Improvements in End-Use Energy Efficiency and Renewable Energy

T. V. Ramachandra

Abstract Energy plays a pivotal role in the development of a region. Increasing dependency on fossil fuels has caused serious concerns at the local (energy dependency, pollution, etc.) and global (global warming, GHG emission, etc.) levels. Harvesting of energy depends on the availability of resources apart from the economic viability and technical feasibility of meeting the demand. The energy requirement of India is mainly supplied by coal and lignite (19378.24 PJ), followed by crude oil and petroleum products (18432.96 PJ) and electricity (7562.24 PJ). However, energy consumption in rural India is largely dependent on non-conventional energy sources due to the availability, possibility of rapid extraction, and appropriate technologies. Globalization and consequent opening up of Indian markets has led to urbanization with the enhanced energy demand in the industrial and infrastructure sectors. The perishing stock of fossil fuel coupled with the growing concerns of climate change has necessitated the exploration of cost effective, environment friendly, and sustainable energy alternatives. Renewable sources of energy such as solar and wind are emerging as viable alternatives to meet the growing energy demand of the burgeoning population. Strengthening of transmission and distribution network with the integration of local generating units (RE-based standalone units) would help in meeting the demand. Distributed generation (DG) with micro grids are required to minimize transmission and distribution (T and D) losses, and optimal harvesting of abundant local resources (such as solar, biofuel, etc.). The focus of the current communication are (i) understanding the energy scenario in India; (ii) sector- and source-wise energy demand with the

T. V. Ramachandra (✉)
Energy & Wetlands Research Group, Centre for Ecological Sciences (CES), Bangalore, Karnataka, India
e-mail: tvr@iisc.ac.in
URL: http://ces.iisc.ernet.in/energy; http://ces.iisc.ernet.in/foss

T. V. Ramachandra
Centre for Sustainable Technologies (Astra), Bangalore, Karnataka 560 012, India

T. V. Ramachandra
Centre for Infrastructure, Sustainable Transportation and Urban Planning (CiSTUP), Indian Institute of Science, Bangalore, Karnataka 560 012, India

S. S. Muthu (ed.), *Energy Footprints of the Energy Sector*, Environmental Footprints and Eco-design of Products and Processes, https://doi.org/10.1007/978-981-13-2457-4_3

scope for energy conservation; and (iii) prospects of renewable energy with smart grids to meet the distributed energy demand while optimizing harvest of local energy sources. Source wise energy analyses reveal that total primary energy consumption has increased manifolds during the past three decades from 18 MTOE (in 1980) to 104 MTOE (2011) in India. Coal consumption has increased from 213 MT (1990–91) to 615 MT (2013–14) and therefore, has grown more than 3 times over the years. Transportation tops in oil consumption (54.28 MTOE) followed by industrial (28.8 MTOE) and domestic (24.89 MTOE) sectors. Total natural gas production in the country was about 18 BCM (billion cubic meters) during 1990–91 and increased to 34.64 BCM now. Electricity generation shows a growth of over 26 times in 40 years that has increased from 43,724 GWh (1970–71) to 11,79,256 GWh. Renewable energy is being used in various forms as is evident from the dependence on bio-energy to an extent of 85% among the rural population (constitutes 70% of the total) since time immemorial. Grid interactive power generating plants from RE sources constitute 37,414 MW with the major share of wind energy plants (24376.26 MW, 65%) followed by biomass/bagasse cogeneration plants (4418.55 MW, 12%), solar photovoltaic (4346.82 MW, 12%), and small hydro (4146.82 MW; 11%). Power generation from municipal solid waste accounts for a very small fraction. Sector-wise and source-wise energy analyses reveal that the energy consumption per GDP (Energy intensity) of India is 0.42 kgoe/million USD. Comparison of the energy intensity (the ratio of energy consumption per GDP) versus GDP per capita of various countries reveal that the energy intensity of India is more than 12 times that of Switzerland, 4 times that of Germany, 3 times that of USA and about 1.3 times that of China, indicating the inefficient use of energy and the need for energy conservation through end use energy efficiency improvements to enhance the GDP with the present level of energy consumption.

Keywords Indian energy scenario · Sustainable energy · Distributed generation Renewable energy · Energy trajectory

1 Introduction

Energy, the basic need of human kind, plays a significant role in the development of a region or country. Energy utilization by human beings has increased from 2,500 kJ/day to more than 2 lakh kJ/day with the evolution of technologies. Every human activity, from crop growing (agriculture) to space research, is dependent on the energy availability and supply. Exploitation of more energy resources helped in innovation of new technologies which made life easier, but caused substantial impacts on the ecology and environment. All economic activities utilize energy. Energy supply has an impact on intermittent production and end use. Economy of the country is influenced by energy, technology improvement from extraction to end use, and supply-demand balance. However, energy is also a limiting factor with inefficient use and fossil fuel dependency (Asafu-Adjaye 2000).

Energy plays an important role in everyday human life and there is disparity in energy consumption across various regions, which depends on the availability, technical, economic, and social aspects. Most parts of India depend on traditional sources of energy such as fuel wood for cooking, water heating, etc. Globally, about 3 billion people depend on bioenergy for domestic purposes and 1.5 billion do not have access to electricity (Rehman et al. 2012; Energy Realities 2013; EIA 2013). Per capita energy consumption varies across countries. It is higher in developed nations (USA—7.3 TOE, Canada—7.6 TOE, Japan—3.7 TOE) compared to the developing (India—0.6 TOE, China—1.8 TOE, Brazil—1.4 TOE) and less developed nations (<0.4 TOE). Figure 1 compares the energy consumption per capita versus GDP (Gross Domestic Product) per capita among the countries (Top 25 GDP countries). Norway (99,933 million USD) tops in GDP per capita followed by Switzerland (79,024 million USD), Australia (65,430 million USD) and Sweden (55,341 million USD) which shows the effective utilization of energy. The per capita GDP value of India is 1555.50 million USD, which is lowest among these countries. Energy consumption per GDP (Energy intensity) of India is higher, hinting the inefficient use of energy. Figure 2 compares the energy intensity (the ratio of energy consumption per GDP) versus GDP per capita of various countries. Energy intensity of India is about 0.42 kgoe/million USD which is more than 12 times that of Switzerland (0.033 kgoe/million USD), more than 4 times that of Germany (0.092 kgoe/million USD), more than 3 times that of USA (0.137 kgoe/million USD) and about 1.3 times that of China (0.325 kgoe/million USD). The prosperity of a nation depends on the efficient use of energy or the energy intensity than the per capita energy consumption.

Most of the Asian countries have high energy intensity (energy/GDP) and lower per capita consumption, which illustrates the inefficient use of energy. This highlights the need of improved end use efficiency to enhance the GDP with the present level of energy consumption (Ramachandra 2011; Ramachandra et al. 2006).

Global studies emphasizing the efficient use of the energy have also demonstrated the relationship between efficient energy consumption and economic

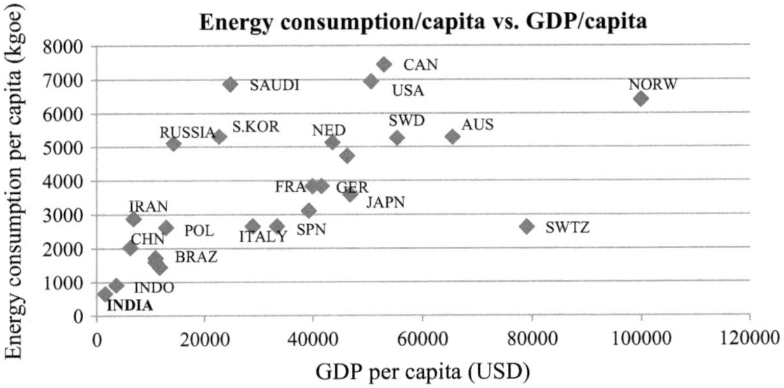

Fig. 1 Country-wise energy consumption per capita versus GDP per capita

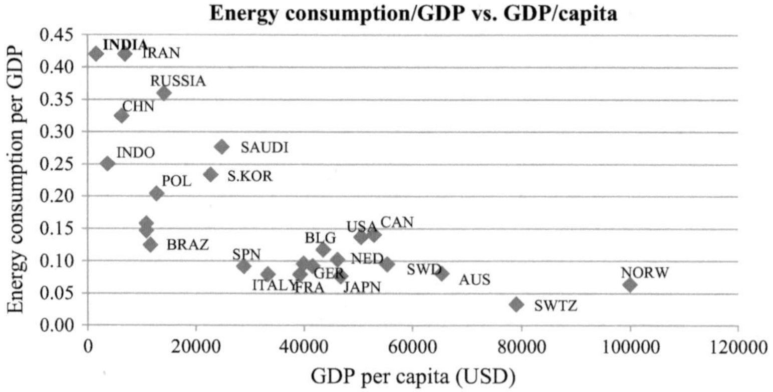

Fig. 2 Country-wise energy consumption per GDP versus GDP per capita

growth. Emission of greenhouse gases (GHG) is proportional to energy utilization and is found higher in developing countries due to the inefficient use of energy (Al-mulali and Che 2012).

Environmental pollution, health related issues, and other global problems have increased with increased fossil fuel extraction and consumption. Unplanned urbanization and industrialization have increased the energy demand. Burning of fossil fuels has led to the emission of greenhouse gases (GHG) such as carbon dioxide (CO_2), oxides of sulfur (SO_x), carbon monoxide (CO), water vapours, etc., apart from the release of particulate matter, solid and liquid waste to the environment. This emphasizes the need for exploiting renewable energy (RE) sources to mitigate pollution and address the problem due to dwindling stock of fossil fuels.

1.1 Indian Energy Scenario

India is the seventh largest geography and ranks fourth among high energy consuming countries in the world with over 1.27 billion population. Total primary energy consumption has increased manifolds during the past three decades from 18 MTOE (in 1980) to 104 MTOE (2011) in India (EIA 2013; TEDDY 2013). Coal, natural gas, and crude oil are the leading commercial sources of energy of the country in which most of the crude oil are being imported. Even though the industrial and commercial sectors make use of fossil fuel resources, the Indian domestic sector largely depends on non-commercial energy sources such as fuel wood, agricultural and horticultural residues, animal residues, biogas, and combustible waste. However, the commercial consumption of bioenergy has decreased with switch over to fossil energy sources (coal, crude oil, natural gas, etc.) over the years (Pachauri and Jiang 2008). Rural population constitutes 70% in India and largely depends on bio resources for domestic energy. About 75% of the rural households depend on firewood, 10% on dung cake, and 5% on LPG for cooking whereas, 22% of the urban households depend on firewood, 22% on kerosene, and 44% on LPG for cooking in the country. Some fraction of the urban households is also dependent on fuel wood for cooking, water heating, and space heating (NSSO 2007). Consumption of non-commercial energy sources in the country remained the same with minimal variations. However, the exploitation of fossil fuels has increased substantially to meet the growing demand of industrial, commercial, and transportation sectors with the favourable policies (Simron et al. 2012). Realizing the growing concerns due to large-scale utilization of fossil fuels on the environment and also to reduce the high imports, India is promoting RE (Renewable Energy)-based energy harvesting programmes through JNNSM (Jawaharlal Nehru National Solar Mission), RGGVY (Rajiv Gandhi Gram Vidyut Yojana), etc., with a goal to have 20,000 MW of grid connected and 2000 MW standalone solar power by 2022. Energy conservation and rural electrification are made mandatory under the Energy Conservation Act, 2011, and Electricity Act, 2003, by the Government of India towards the goal of energy independence and to lower the energy demand (MNRE 2013a). Economic development with financial security of a region is dependent on energy independence and achieving the same is a daunting challenge. Exploitation of renewable sources with efficient use of energy and demand-side management would ensure sustainable growth in the energy sector while reducing environmental pollution (Bhattacharyya 2010). This will bridge the supply-demand gap through reduction in the energy loss from generation to end use.

Electricity has a wide range of applications as it is a clean and an efficient media of energy transport. Per capita electric energy consumption in India is about 879 kWh (2012) and the source of electricity generation plays a significant role in energy management and conservation. Electricity generation has been largely dependent on fossil fuels (coal) which are mostly centralized. Centralized generation and sparsely located loads are the prime reasons for un-electrified rural households with higher transmission and distribution (T & D) losses. Indian electrical power transmission and distribution network encounters higher losses ($\sim$24%) compared to other countries (China—6%, Australia—5%, Bangladesh—10%, Germany—4%) and world average ($\sim$10%) due to un-metered electricity supply, un-authorized expansion, theft and pilferage at the distribution side (CEA 2013). Transmission networks are being strengthened with the advanced (electronic) metering facility in many urban regions. In this context, innovations in power sector through distributed/decentralized generation (DG), micro-grid and smart grid would pave the way for efficient and effective power systems in India (ISGTF 2013; MoP 2013).

Implementation of DG results in direct economic benefits including reduction in operation and maintenance (O&M) costs, capital investment to upgrade the generation, fuel cost, and dependency on fossil fuels. Other indirect benefits include reduced investment for pollution prevention and health-related expenses, apart from achieving the national and local energy independence (El-Khattam et al. 2005; Pathomthat and Ramakumar 2004). DG and micro-grid have the versatility of incorporation of next generation power technologies such as smart grid architecture, advanced metering infrastructure (AMI), biofuel generation using algae, electricity from solid waste, and energy plantation. Certainly the future energy resources have to renewable in nature, to meet the growing demand; distributed generation and micro-grid facilitates energy generation near load centers. Exploitation of RE sources and reduction of T & D losses certainly make the energy sector sustainable while achieving energy independence.

1.2 Need for the Study

India with fastest growing economy, the dependence of energy has increased manifold due to industrialization and the impetus given to infrastructure development. The trajectory of energy generation, transportation and consumption has to be understood in order to adopt the sustainable energy management strategies to avert any energy crisis (Ramachandra 2008; Ravindranath and Balachandra 2009). Effective DSM (Demand Side Management) techniques include budgeting non-commercial energy resources in the present energy scenario and end use efficiency improvements. The present study analyses developments in the energy sector in India from generation to end use. Energy conservation needs to be achieved through the improvements in the end-use devices to have a sustainable and pollution free growth. Technological interventions will give scope to utilize locally available non-conventional, renewable energy resources (Kumar et al. 2013). Substitution of fossil fuels through RE sources will also help in attaining energy independence in the region. However, this requires assessment of the resources and patterns of energy consumption from various resources (commercial and non-commercial), apart from techno-economic feasibility of alternate energy trajectory.

Centralized options of generating electricity and supply to remote and sparsely located loads have often faced technical challenges apart from lack of economic viability. Electrification of rural India is yet to gain momentum as is evident from the absence of electricity supply to more than 74,00,000 households of 18,000 villages. The electrification of remote villages to meet the basic electricity demand is possible through standalone RE source based generation (Ramachandra and Shruthi 2005; Ramachandra et al. 2014a; Nouni et al. 2009). Potential assessment of RE sources through geographical information system (GIS) has helped in optimizing the availability (Ramachandra et al. 2014b) and integration of resources towards a reliable supply. Energy conservation and demand side management is the other aspect to flatten the load curve and to reduce the peak demand. DSM techniques and efficient methods in end use of energy will have a direct impact on generation. The present study analyses the prospects of RE sources, non-commercial energy sources, scope for the improvement of end use efficiencies, and options to ensure sustainable energy path, while reducing the supply-demand gap. The main objectives of the current research are as follows:

(i) Understanding the energy situation in India—sector- and source-wise energy demand with the scope for energy conservation through DSM and end use efficiency,
(ii) Prospects of renewable energy with DG and smart grids to meet the distributed energy demand while optimizing harvest of local energy sources.

This chapter consists of three major section—second section brings out the source-wise energy scenario over the last five decades. Third section elucidates the scope for renewable energy and the fourth section argues for the energy conservation through improvements in end use efficiencies, smart grid framework, etc.

2 Temporal Changes in Energy Utilization: An Overview

Energy resources can be categorized as commercial (coal, crude oil, natural gas, etc.) and non-commercial (fuel wood, animal residues, agricultural and horticultural residues) depending on the market mechanism. Commercial energy resources are mainly used in industries, transportation, electricity generation, and commercial sectors. Domestic sectors' energy demand is mainly met by locally and freely available non-commercial resources in rural areas (Ramachandra et al. 2000). India is blessed with very few energy resources such as coal, natural gas, biomass, water, etc. Traces of crude oil sources are found in few places which are not able supply the huge demand. Analysing the temporal change of energy resources will help to understand the resource status and take appropriate action to meet the forthcoming demand.

2.1 Coal

Coal mining in India started during the eighteenth century for meeting the fuel demand of railways and industries. India is the third highest coal producer in the world with annual production of 739.92 million tonnes (MT) in 2013–14. India has a proven coal reserve of 293.5 billion tonnes available at Jharkhand, Chhattisgarh, Odisha, Madhya Pradesh, and West Bengal (MoC 2013). Resource assessment predicts that the approximate life of coal is 169 years with present reserve and consumption. Figure 3 gives the temporal patterns in coal production and imports over the last two decades. Coal consumption has increased from 213 MT (1990–91) to 615 MT (2013–14) and therefore, has grown more than 3 times over the years. About 168.44 MT of coal is imported in 2013–14 to meet the growing demand in the power sector.

Calorific value of Indian coal varies from 4000 to 7000 kcal/kg, which is relatively lower. Coal utilization for electric power generation tops the consumption with 463.71 MT (2013–14) followed by the industrial sector (Fig. 3). Steel industries consume 23.16 MT, followed by cement (13.36 MT) and textile industries.

There are numerous environmental and economic problems associated with coal mining, combustion, and power generation. Coal mining results in geographical

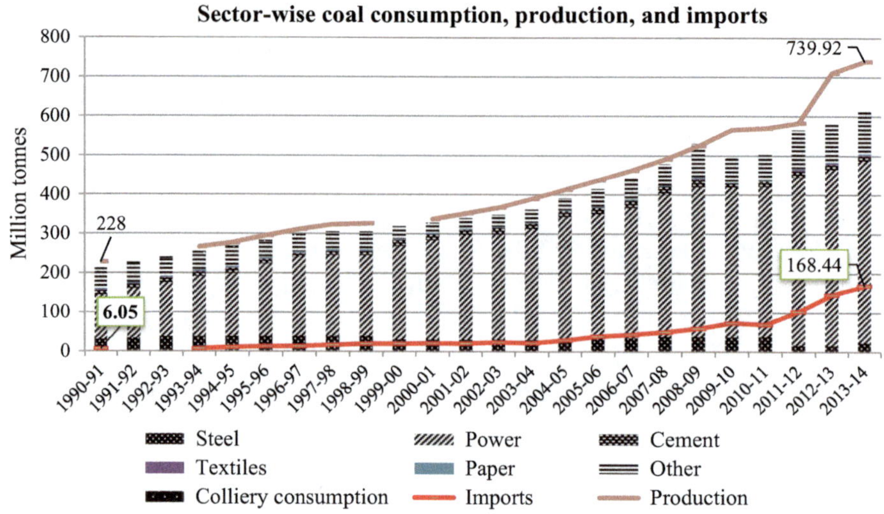

Fig. 3 Temporal change in coal production, imports, and sector-wise consumption

change of land, with the minimal scope for restoration to its original state. Coal-based thermal power plants have been polluting the neighbouring environment, thus, causing respiratory and other health problems (Ramachandra et al. 2012). Air pollution from coal mines is mainly due to the emissions of particulate matter and greenhouse gases (GHG) including methane (CH_4), sulfur dioxide (SO_2), nitrogen oxides (NOx), carbon monoxide (CO), etc. These pollutants contaminate water resources, air, soil, and consequent changes in the climate.

2.2 *Crude Oil, Natural Gas and Petroleum Products*

Transportation tops in oil consumption (54.28 MTOE) followed by industrial (28.8 MTOE) and domestic (24.89 MTOE) sectors. Figure 4 gives the sector-wise consumption of crude oil in the country (2012–13). India is blessed with very few oil resources which are located in Mumbai High, Bay of Bengal, and Rajasthan which are being monitored by the Oil and Natural Gas Corporation (ONGC).

The total oil production in India has marginally increased from 33 million tonnes (1990–91) to about 37.7 million tonnes (2013–14). Consumption of oil has increased multiple times with industrialization and revolution in transportation system, leading to a radical escalation in oil imports. Imports of crude oil increased from 20.7 million tonnes (1990–91) to 189 million tonnes (2013–14). Figure 5 illustrates the temporal changes in crude oil production and imports (MoPNG 2013). Drastic change in imports occurred during the end of last century, emphasizing the need for energy conservation, demand side management, and renewable energy-based capacity addition, in order to minimize the dependency on other countries.

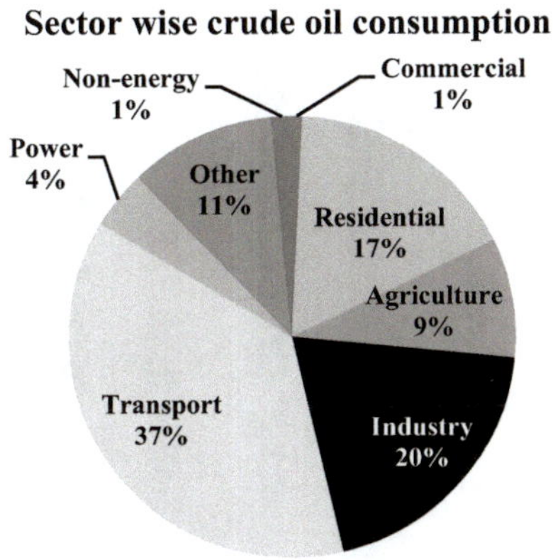

Fig. 4 Sector-wise crude oil consumption (MTOE) in India

Natural gas is one of the prominent energy sources in power and industrial sectors. Total natural gas production in the country was about 18 BCM (billion cubic meters) during 1990–91 and increased to 34.64 BCM in 2013–14. Natural gas is available in India at Mumbai High basin and in Gujarat. Offshore gas reserves are also located in Andhra Pradesh coast (Krishna Godavari Basin) and Tamil Nadu coast (Cauvery Basin). Onshore reserves are located in Gujarat and the North Eastern states (Assam and Tripura). The country has 1437 BCM of natural gas reserves which may last for 30 years at the present level of consumption. Figure 5 depicts the natural gas production and annual imports from 1990–91 to 2013–14.

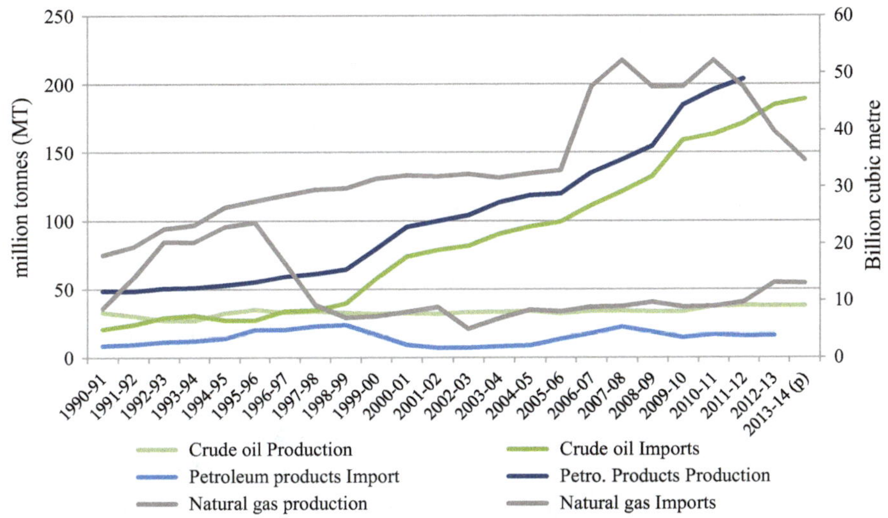

Fig. 5 Temporal change in crude oil, natural gas, petroleum products, and their imports

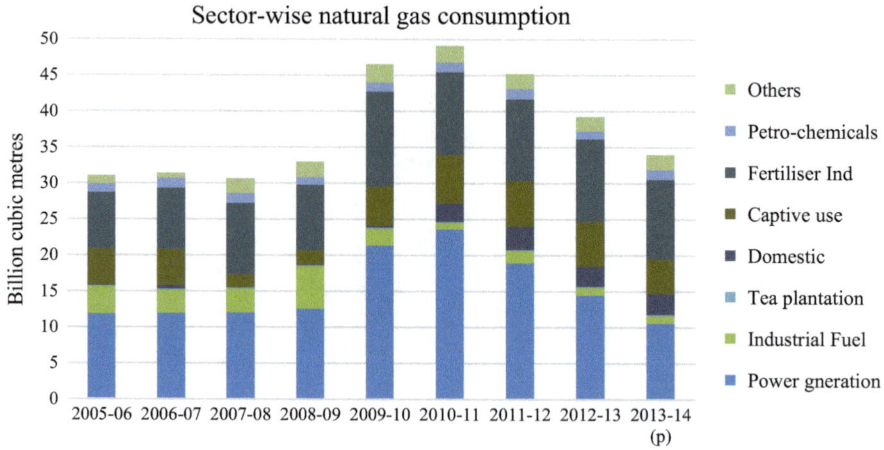

Fig. 6 Sector-wise consumption of natural gas in India

Figure 6 shows the sector-wise consumption of natural gas in the country. Fertilizer industries are the major consumer of natural gas (11.06 BCM) followed by power generation (10.53 BCM) (MoPNG 2013). Petroleum products and their utilization have been increasing at a higher rate since globalization era (after 1990s). The total consumption of petroleum products was about 55 MT (1990–91) which has increased to 148 MT (2011–12). Figure 5 also gives the temporal change in petroleum product production and consumption. Import of petroleum products is about 15.85 MT (during 2011–12) and the increase in crude oil imports is directly dependent on the rise in petroleum product consumption. However, the import of final products is expensive compared to crude oil, which affects the country's economy. Production of petroleum products also increased significantly from 48.5 MT (1990–91) to 204 MT (2011–12), indicating the four-fold growth (MoPNG 2013). Rapid increase in consumption has resulted in numerous environmental and economic problems due to inefficient combustion. This necessitates a paradigm shift from fossil fuel-based energy to renewable energy to achieve the energy independent sustainable development.

2.3 *Electric Power*

India is one of the major electric energy consuming countries in the world with the annual generation of 11,79,256 GWh (2013–14) and largely depends upon fossil energy resources. Figure 7 illustrates the share of energy sources in total installed power capacity in the country. Coal is the prominent energy source (170,737.88 MW) followed by hydro (42,623.42 MW), nuclear power plants (37,415.53 MW) and renewable energy sources (27,541 MW). The government

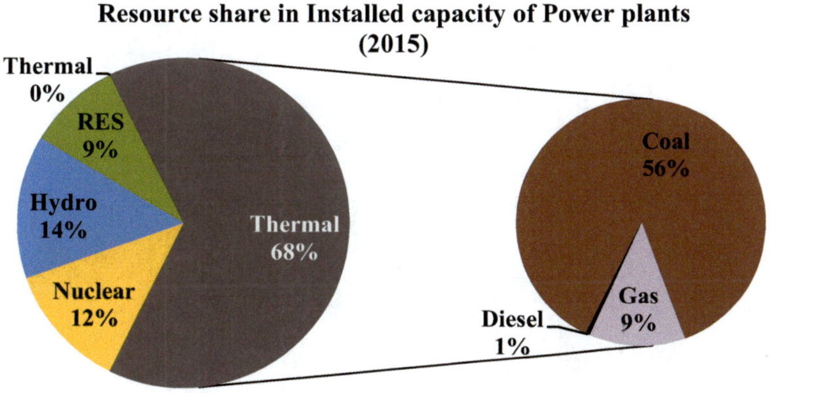

Fig. 7 Share of energy sources in total installed capacity

has proposed augmentation of nuclear power generations to 20,000 MW by 2020 (CEA-LGBR 2013). Electricity generation has increased from 43,724 GWh (1970–71) to 11,79,256 GWh (2013–14) showing a growth of over 26 times in 40 years. Industrialization, revolution in agriculture and elevated consumption in commercial sector have necessitated an additional increment in electricity demand. Figure 8 gives the sector wise temporal electric energy consumption. Industries top the consumption with 346,469 MW (44.8%), followed by agriculture (133,650 MW, 17.3%), domestic (170,034, 22%) and commercial sector (69,266, 9%) (MoP 2013).

2.4 *Renewable Energy (RE) Sources*

India is bestowed with ample renewable energy resources throughout the region with the scope for harvesting wind, bio-energy, solar (PV and Thermal), micro hydro, biogas, geothermal, tidal energy, etc. (Ramachandra 2011). Figure 9 depicts the

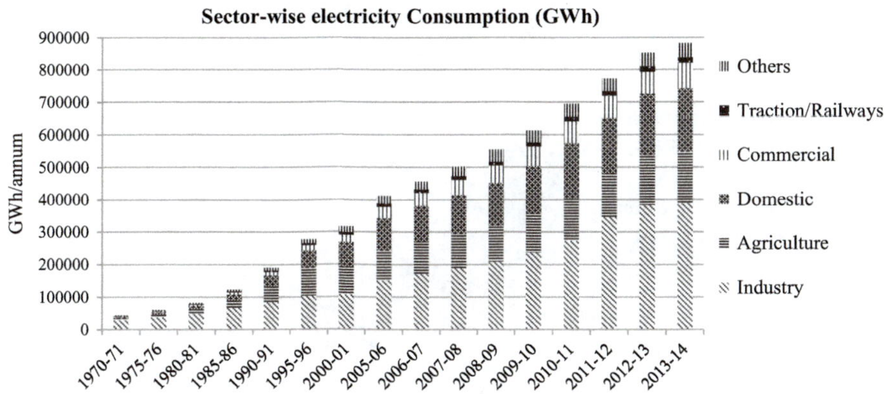

Fig. 8 Sector-wise electricity consumption in India

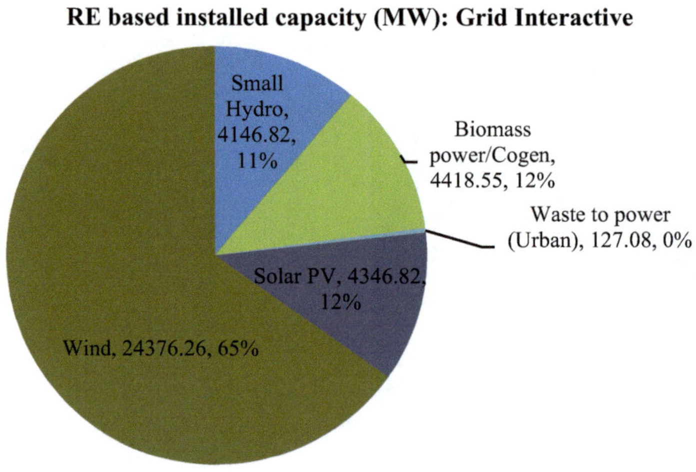

Fig. 9 Share of RE sources (Grid interactive) in total installed capacity

share of RE sources, based on the total installed capacity. Renewable energy is being used in various forms as is evident from the dependence on bio-energy to an extent of 85% among the rural population (constitutes 70% of the total) since time immemorial. Grid interactive power generating plants from RE sources constitute 37,414 MW with the major share of wind energy plants (24376.26 MW, 65%) followed by biomass/bagasse cogeneration plants (4418.55 MW, 12%), solar photovoltaic (4346.82 MW, 12%), and small hydro (4146.82 MW; 11%). Power generation from municipal solid waste accounts for a very small fraction (MNRE 2013a, b).

Figure 10 shows the share of RE energy sources in off-grid installation (1228.48 MW_e) for remote area electrification and as captive generation in industries.

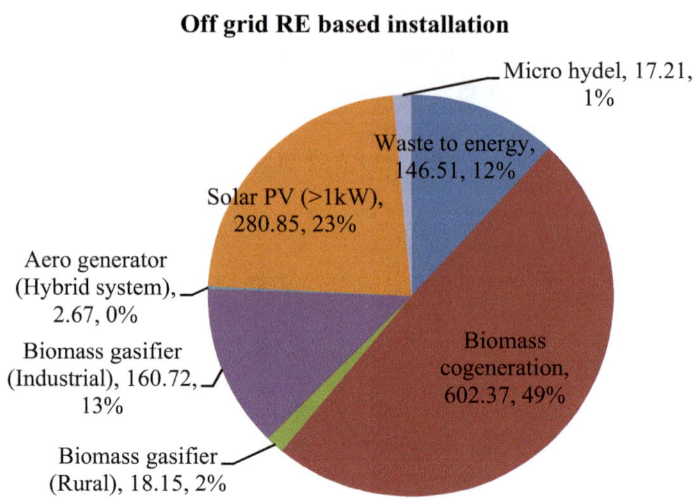

Fig. 10 Share of RE energy sources (MW_e) in off-grid installation

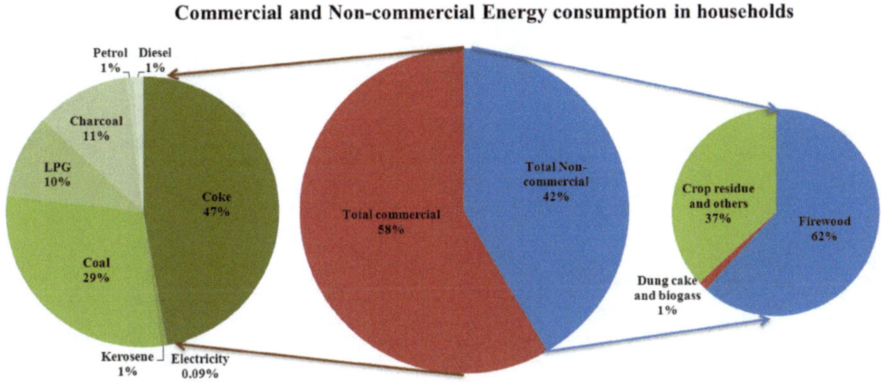

Fig. 11 Commercial and non-commercial sources in domestic energy consumption

The Government of India is encouraging RE-based capacity installation through National Solar Mission, Jawaharlal Nehru National Solar Mission (JNNSM), projects under Clean Development Mechanism (CDM), incentives for biogas installation, etc. Policy initiatives such as Feed-in-Tariff (FIT) and Generation Based Incentives (GBI), tax holidays, and subsidy on capital investments will help to boost the RE-based capacity addition. Sustainable energy development can take place with renewable energy-based generation while reducing the GHG emission.

Solar cookers, dryers, and improved cook stoves can be used in the domestic sector whereas, solar and wind driven pumps are reliable in irrigation (Ramachandra 2011). Captive electric energy generation using solid waste, bagasse, agricultural and horticultural residues, wind and solar energy are viable in the industrial sector. Hence, the RE sources can replace the present energy mix with a higher share with distributed generation and micro-grid (rooftop) generation.

The Indian energy scenario shows that the present energy mix is dominated by conventional energy sources. Dependency on fossil fuels has increased the imports which is affecting the country's economy. Development in the energy sector is likely to deviate from sustainable path hinting the energy crisis in future. Figure 11 gives the share of commercial and non-commercial sources in the energy mix of the domestic sector of the country. About 58% of the demand is met by commercial energy sources wherein coke dominates the consumption (47%), followed by coal (29%), charcoal (11%), and LPG (10%). Non-commercial sources supply 42% of the demand in which firewood tops with a share of 62%, followed by crop residues and others (37%). Figure 13 highlights an increase in carbon dioxide (CO_2) emission from 143 million tonnes (1962) to 2,073 million tonnes (2014). Mitigation of changes in the climate entails lowering GHG emissions, through energy conservation and improvements in end use devices efficiency.

The current level of energy extraction, transportation, and inefficient consumption pattern and its impact on the environment and nation's economy has necessitated a paradigm shift in the energy planning to achieve sustainable development.

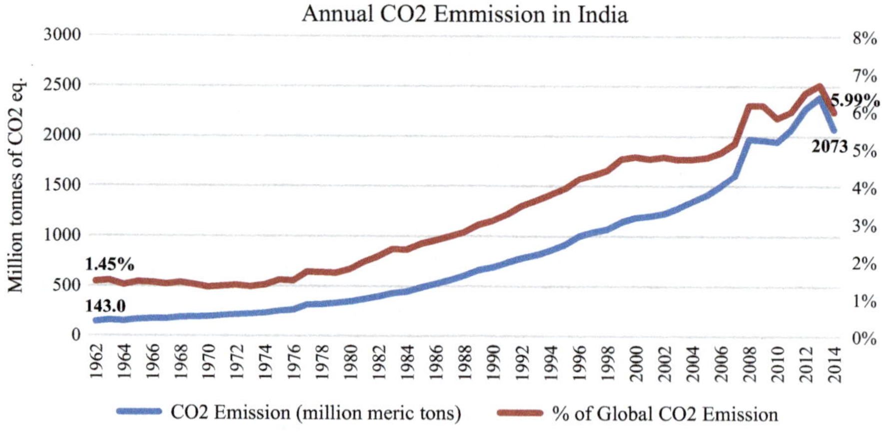

Fig. 12 CO_2 emission (million tons) from 1962 to 2014

Figure 12 shows the increasing trend of GHG emissions. Power generation and transportation sectors are the major contributors to emissions which entails the transition to renewable energy and innovations in vehicle design, etc. Apart from that, end use energy efficiency improvement with DSM would help to conserve energy. Adoption of smart grid architecture would further improve the energy sector through intelligent, reliable, efficient, and less pollutant systems (Ramachandra 2009a; Dabrase and Ramachandra 1999; Prakash and Bhat 2009).

3 Scope for Renewable Energy

3.1 Solar Energy

India is one of the best recipients of solar energy due to its favourable location in the solar belt (40 °S to 40 °N) and receives annual sunshine of 2600 to 3200 h.

Figure 13 illustrates that the Gangetic plains (Trans, Middle, and Upper) Plateau region (Central, Western, and Southern), Western dry region, Gujarat Plains, and hill region as well as the West Coast plains and Ghat region receives annual global insolation above 5kWh/m^2/day. These zones include major federal states of Karnataka, Gujarat, Andhra Pradesh, Maharashtra, Madhya Pradesh, Rajasthan, Tamil Nadu, Haryana, Punjab, Kerala, Bihar, Uttar Pradesh, and Chattisgarh. The eastern part of Ladakh region (Jammu & Kashmir) and minor parts of Himachal Pradesh, Uttarakhand, and Sikkim, which are located in the Himalayan belt also receive similar average global insolation annually. These regions with a viable potential constitute solar hotspots covering nearly 1.89 million km^2 ($\sim$58%) of India (Fig. 14) with the favourable prospects for solar-based renewable energy technologies, which could help meet her escalating power requirements in a

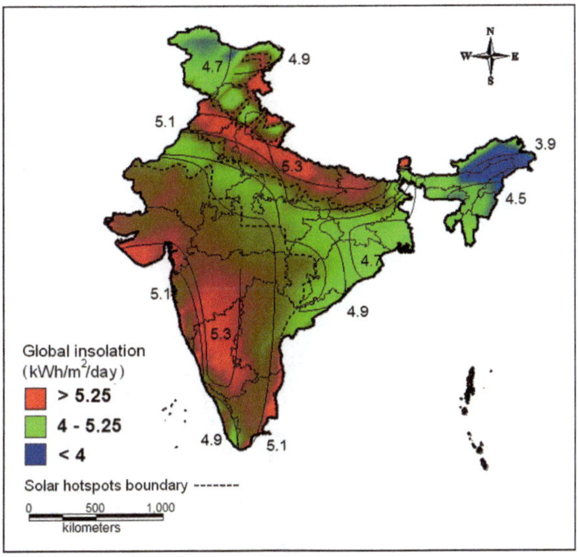

Fig. 13 Annual average global insolation map of India showing the isohels and solar hotspots

decentralized, efficient, and sustainable manner. A techno-economic analysis of the solar power technologies and a prospective minimal utilization of the land available within these solar hotspots demonstrate their immense power generation as well as emission reduction potential. A major thrust for R&D in solar technologies is essential to lower the generation cost and enable competition with the conventional fossil fuel-based options.

Regions receiving global insolation of 5kWh/m^2/day and above can generate at least 77 W/m^2 (actual onsite output) at 16% efficiency. Hence, even 0.1% of the land area of the identified solar hotspots (1897.55 km^2) could deliver nearly 146 GW of SPV-based electricity (379 billion units (kWh) considering 2600 sunshine hours annually). Figure 14 gives the district wise solar power density of the country. This power generation capacity would enhance considerably with the improvement in efficiency of SPV technology. Solar technologies have the potential to offset a huge volume of GHG emissions as demonstrated and help realize a low carbon economy at a faster rate. It will create numerous employment opportunities, especially at the village level. Learning from other developing countries as well as its own past experiences, India can be a world leader in solar power generation. With an ambitious solar mission, and positively evolving policy instruments, the nation will rightly adorn the epithet of 'Solar India' in the near future.

The National Solar Mission (NSM) launched in 2009 by the Government of India has given a great boost to the solar scenario in the country. The mission targets achieve 175 GW by the year 2022 which includes 100 GW from solar, 60 GW from wind, 10 GW from bio-power, and 5 GW from small hydro-power. The solar energy installation comprises of 40 GW rooftop and 60 GW through large- and medium-scale grid connected solar power projects. About INR 6,00,000 crores

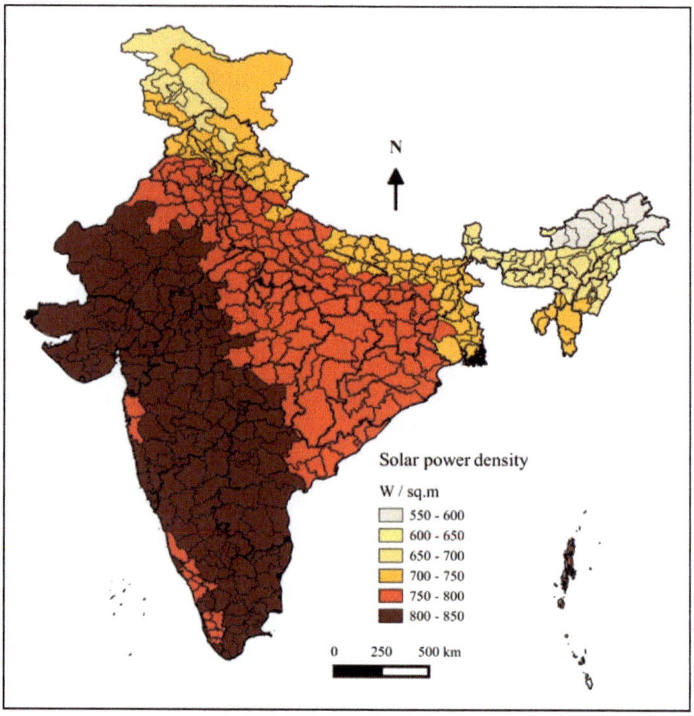

Fig. 14 District-wise solar power density of the country

investment is expected for the commission of 100 GW solar projects (CEA—LGBR 2013). However, considering the current level of T & D losses in a centralized system, inefficient, and unreliable electricity supply, it is necessary to promote decentralized energy generation. Small capacity systems are efficient, economical, and more importantly would meet the local electricity demand. The incentives could be

(i) Solar Rooftop PV systems can be installed on residential/commercial/industrial buildings in the state. Excess generated energy can be fed to the grid with net metering with incentives (of INR 9.56/unit—without subsidy and INR 7.20/—with subsidy).
(ii) Buyback programmes for the electricity generated at household level and in micro grid—GBI of INR 9.56 for electricity generation (<5 kW) feeding to the grid by SPV.
(iii) Install solar rooftops in all new government/local body buildings—implementation of solar rooftops could be in a phased manner in the existing government/local body buildings, etc.
(iv) Commercial lighting in advertisement boards should only be from RE sources. Complete ban on usage of grid electricity for these purposes.

(v) Impetus to energy research through generous funding for the R & D activities to ensure further improvements in the grid, technologies, two-way communication energy meters (to connect rooftop generation with existing grid), efficient luminaries' production, low cost wiring, switchgears, appliances, etc.
(vi) Energy education (focusing mainly on renewable energy technologies, end-use energy efficiency improvements, energy conservation) at all levels. School curriculum shall include renewable energy (RE) concepts.
(vii) Awareness about energy independence and the necessity of RE sources in the present gloomy energy scenario to the consumers.
(viii) Education and awareness about applications and importance of renewable energy sources.
(ix) Capacity building of youth through technical education for installation and servicing of SPV panels.
(x) Setting up service centers in block development offices to meet the requirement of service support for RE technologies (solar, biogas, energy efficient chulas, etc.).

3.2 *Wind Energy*

India ranks fifth (after China, US, Germany, and Spain) with over 19 GW wind installed capacity. Wind energy accounts for 8.5% of the total installed capacity. Figure 15 gives the district-wise wind power density potential in India. The total wind energy potential in country is estimated as 49.13 GW in which about 38% has been utilized for energy generation (Sharma et al. 2012). The state of Tamil Nadu leads in wind energy extraction with the installed capacity of 6,286 MW followed by Maharashtra (2,400 MW), Gujarat (2,337 MW), and Karnataka (1,773 MW) (C-WET 2012). Energy extraction from wind resources primarily depends on the wind speed available in the region. The available wind energy potential is directly proportional to the wind speed and area swept by the wind turbine. Hence, the primary need is to assess the annual wind speed of the region which indicates the potential regions for energy extraction. The coastal region of the country experiences high wind speed which ranges from 3 to 5 m/s annually. The southern and central part (west coast) of the country experiences higher wind speed during monsoon (June to September) which will be more than 5 m/s. During winter, the high elevated region of the country experiences high flow of wind that ranges from 4 to 5.25 m/s (Ramachandra and Shruthi 2007). Estimation shows that the western coast (Karnataka, Tamil Nadu, Kerala, Maharashtra, Gujarat) and plains (Rajasthan, Gujarat, Karnataka) are the ideal places for wind energy harvesting where the annual average wind speed is higher. However, there is vital scope for decentralized wind energy generation in hilly regions (such as Jammu and Kashmir, Himachal Pradesh) and islands (such as Anadaman and Nicobar, etc.). Distributed wind

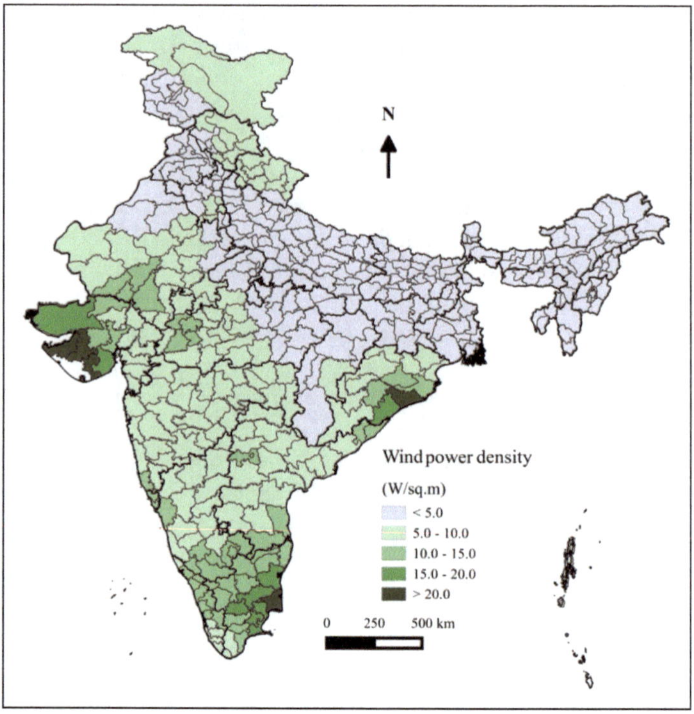

Fig. 15 District-wise wind power density potential in India

applications in water pumping and milling could meet the energy demand for irrigation and domestic sector of the region (Ramachandra and Krishnadas 2012).

3.3 Bio-Energy

Bio-energy is a prominent component of total primary energy consumption in India. About 70% of the population lives in the rural region of the country where agriculture and horticulture are the primary occupations. Residues obtained during the processing of the yield are one of the major sources of bio-energy in the country (Ramachandra et al. 2014). Forest residues and fuel wood are the primary energy sources for heating and cooking in rural India. The sector-wise available bio-energy is estimated and compared with the energy demand. Figure 16 gives the state-wise supply to demand ratio of biogas and biomass energy, which shows the ratio of 0.25–0.5 for most of the states. Cattle dung and biogas generation meets the significant domestic energy demand for cooking in rural and suburban areas. The north-eastern part of the country with good forest cover shows a better resource status.

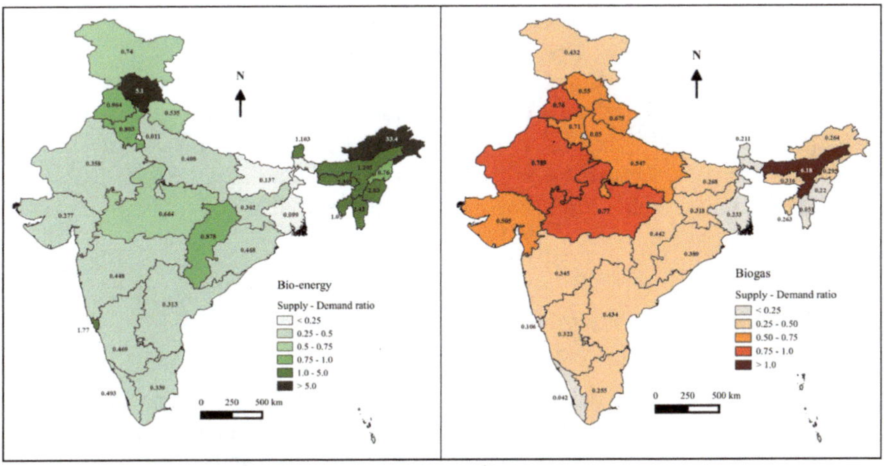

Fig. 16 State-wise bioenergy and biogas status (supply/demand ratio)

The Ministry of New and Renewable Energy (MNRE), Government of India, has come up with policies and plans to encourage bio-energy utilization in the country. The country has a total installed capacity of 1,284 MW biomass power plants and 2,392 MW of bagasse generation plants which are synchronized with grid (MNRE 2013b). Waste to power generation grid interactive plants of about 100 MW installed capacity demonstrate the productive way of municipal solid waste (MSW) management (MNRE 2013a; Ramachandra 2009b). The country has more than 750 MW installed capacity off grid/captive power plants which gives the new avenues for decentralized power generation (MNRE 2013b). Improvement in bio-energy technologies (BETs) and effective management of wasteland, friendly policies, and incentives would certainly increase the energy potential and capacity addition using more bio-energy resources (Singh and Setiawan 2013).

3.4 *Biofuel*

In the face of increasing CO_2 emissions from conventional energy sources and the projected scarcity of crude oil, there is an immediate need for cost effective renewable alternative energy sources. Bio-diesel generation from gasoline secreting diatom solar panels is a revolutionary change to meet the future crude oil demand. Diatoms, the major group of planktonic algae, can be used sustainably for production of bio-fuel, by the usage of diatom-based solar panels. Studies have shown that diatoms could make 10 to 200 times as much oil per hectare as oil seeds. Some diatoms secrete more lipid content when subjected to unfavourable environment or culture conditions, such as nutrient starvation or extreme temperatures (Mahapatra et al. 2014). Since diatoms multiply rapidly, they can double their biomass within an hour to a day's time. Since each diatom creates and uses its own gas tank, it is estimated that diatoms are responsible for up to 25% of global carbon dioxide (CO_2) fixation. This shows that while diatoms can be cultivated for oil extraction, they can automatically reabsorb carbon dioxide in the process. Diatoms have the potential to meet the future oil demand which also plays a major role in CO_2 absorption. This enables the scope for mimicking the natural process to extract oil which leads to sustainable growth (Ramachandra et al. 2013).

3.5 Biofuel from Wastewater Algae

Third generation biofuel, based on microalgae, is emerging as one of the most promising sources due to algae's high photosynthetic efficiency and faster replication as compared to other energy crops. However, optimization of the conditions for the growth and technologies for biomass harvest and energy extraction are necessary for sustainability, together with a cost effective way of algal cultivation. Abundant wastewaters, generated in urban localities every day, provides the nourishment to nurture algae for biofuel generation. Domestic wastewaters potentially provide economic and sustainable means of dense algal growth. Algae have the ability to uptake nutrients which aid in the treatment of wastewater. The total lipid content of *Euglena* species was higher (24.6%) compared to *Spirogyra* sp. (18.4%) followed by *Phormidium* sp. (8.8%) and their annual lipid yield potential was 6.52, 1.94, and 2.856 t/ha/yr., respectively. These species showed higher content of fatty acids (palmitate, stearate followed by oleic and linoleic acids) with the desirable biofuel properties. This suggests that algae based treatment option for removal of nutrients from wastewater as well as biofuel production for fostering the sustainable production of renewable energy. Thus, extraction of lipid from micro-algae, grown in wastewater, would serve the dual purpose of cost effective waste treatment and help in meeting the regional energy demand (Ramachandra et al. 2009).

3.6 Capacity Addition Through Renewable Energy Sources

The Indian power sector is facing installed capacity deficiency problem due to the ever-increasing load. A large number of new loads are being added to the grid, but increasing installed capacity is not an overnight process. Accumulated load has severe impact on the power system supply which is a challenging task. The present generating stations are working to their maximum capacity and most of them are centralized. Power sector equipment (transformers, transmission lines, insulators, compensators, etc.) are aged, working with lower efficiency; replacing or up-gradation is a costly affair and takes more time. Overloading of such equipment is presently not possible which may lead to blackout. Adoption of new trending technologies such as smart grid, energy management system (EMS), SCADA is a tough task and expensive with the present power system network. Connecting un-electrified load to the present grid increases the load which might collapse the grid. In this perspective, to meet the ever growing load, there is a need to exploit the renewable energy potential in the country. Capacity addition through RE sources and decentralized installation of generation plants will reduce the load on transmission network and also narrow the energy demand gap (Nouni et al. 2008; Hiremath et al. 2007).

Renewable energy technologies (RET) such as individual/community level rooftop installation of solar PV, biomass gasifiers, wind energy conversion systems, biogas plants, etc., have the potential to substitute grid electricity. Since the country receives solar insolation over 5 kWh/m^2/day for more than 300 days annually, solar PV installation on rooftop and in wasteland could be viable option to build up the capacity. India has over 7,000 km of coastline which are high potential wind regions. Installation of wind turbine near sea shores (fraction of area) could generate enormous amount of energy which also adds to the natural splendour. Most of the Indian population residing in rural areas practices agriculture. Agricultural and horticultural residues have the potential to meet village level domestic energy demand through gasification. The prime advantage of this system is that it produces electricity, gas, and manure which can be returned to the farmer. These systems can be installed by individuals or as a community (pay for service) in larger scale which can also be connected to the grid. Hybridization of locally available RE sources makes the system more reliable, efficient, economically viable, and sustainable (Ghosh et al. 2002; Balamurugan et al. 2009).

4 Energy Conservation and New Energy Technologies

4.1 End-Use Efficiency Improvement

More than 70% of the population resides in rural regions and 85% of the energy requirement is met by traditional fuel through energy inefficient devices. Industrial energy consumption is also inefficient in most of the cases due to the aged equipment, lack of lubrication, torn out parts, and non-scientific combustion. The overuse of energy resources in the commercial domain and unmetered energy supply for irrigation pumps have aggravated the energy crisis.

The primary need of energy resources in rural India is for cooking, water/space heating, and lighting. Most of the energy for cooking and heating is supplied by bioenergy (fuel wood, dung cake, etc.) which is locally available. However, the conventional cook stoves used for combustion of biomass have lower thermal efficiency (<10%). Compared to these, improved cook stoves (ICS) have higher efficiency (20–30%) and there is a scope to reduce 27 to 42% of the fuel wood requirement (Ramachandra et al. 1999). A typical rural household consumes about

5 l of kerosene every month. Average electricity consumption in rural household ranges between 50 and 60 kWh/month which is mainly used for lighting, entertainment, water pumping, and air cooling. About 30–40% of energy conservation is possible in the domestic sector using CFL/LED lamps for lighting, energy efficient heaters, and coolers (Reddy 1999).

The domestic energy requirement of an urban household is supplied by electricity, LPG (Liquefied Petroleum Gas), fuelwood, etc. Even though an urban household consumes about 11 kg of LPG per month, 22% of the urban households depend on firewood and kerosene as primary energy need. Electricity is the main source of lighting, cooling, and water heating in urban area where the consumption ranges from 100 to 125 kWh per month (TEDDY 2013). Use of ICs, CFL/LED lamps, and energy efficient heaters and coolers can conserve a significant amount of energy. Solar water heater and rooftop solar PV installation can substitute electricity and biomass consumption for lighting and water heating, respectively (Vishwanathan and Ravikumar 2005).

Energy conservation in irrigation pump sets is possible by avoiding over capacity installation, maintenance and lubrication, selecting proper foot valves and pipelines, drip irrigation, and sprinkler installation, etc. Energy supply for agricultural purposes is to be metered and tariff has to be applied on the basis of installed capacity. This would help in the optimal irrigation of agriculture fields. Wind pumps and solar PV pumps can be installed for small area irrigation (5–10 hp) which would replace the diesel or kerosene fueled pumps (Kumar et al. 2010).

Industries are the highest energy consumers in India which use all forms of energy resources. Many of the Indian industries use coal, oil, and electricity. About 30–40% of energy conservation is possible with upgradation of equipment and technology. However, there is a need to reform policies and tariffs for industrial energy consumption to promote captive generation through renewable energy sources (Gupta and Sengupta 2012; Ramachandra and Subramanian 1995).

Energy consumption in the commercial sector has increased considerably during the last decade. Energy conservation in the commercial sector through interventions in lighting technologies (LED/CFL), green buildings, and energy efficient equipment would reduce the energy consumption and decrease the energy intensity.

4.2 Demand Side Management (DSM)

The techniques and measures taken in load side to improve the reliability and quality of power are termed as demand side management (DSM). DSM techniques include use of energy efficient equipment (CFL/LED lamps), reactive power compensators (STATCOM, series/parallel capacitors/inductors), load shifting, and load shaving, etc. (Palensky and Dietrich 2011). Capacity addition through renewable energy sources is also a DSM technique which reduces the consumer's dependency on the grid. DSM techniques immediately affect the power system (generation and load) which narrows the supply-demand gap. Demand response, the widely used DSM technique, basically includes two ways—Intensive based programme (IBP) and Price-based programme (PBP). In IBP, consumer will not derive direct benefit for cutting down the load, however, the same shall be earned through incentives such as tax reduction and tax holidays. In PBP, different types of power tariffs are applied to the consumer, from which direct benefit is possible in electricity bills. Various tariffs such as time of use (TOU), maximum demand (MD) pricing, critical peak pricing (CPP), real time pricing (RTP), power factor tariff, etc., will control the electricity bill (Albadi and El-Saadany 2007).

4.3 Smart Grid and Energy Management System

Smart grid is an intelligent system (manual/automated) which integrates all components of the power system (generator, transmission and distribution network, end users) for reliable, efficient, and environment-friendly energy supply. It also plays a key role in demand response, peak load management, unit commitment, and to have effective renewable mix in the installed capacity. Well-established information and communication technology (ICT) and control networks are the backbone of smart grid for which the supportive grid network is required (Vijayapriya and Kothari 2011). Power sector in India is evolving and adopting modern grid technologies such as supervisory control and data acquisition (SCADA), energy management system (EMS), distribution automation (DA), advanced metering infrastructure (AMI) such as prepaid meters, etc. However, the communication network is limited to high voltage transmission equipment and feeble parts of the present power network need to be strengthened to have the smart grid architecture. India is planning to have a full phase smart grid by 2025, for which required devices such as FACT (flexible AC transmission) controllers and phasor measurements units (PMUs) are being installed. Around 14 pilot projects are being implemented by Indian Government under Restructured Accelerated Power Development and Reforms Programme (R-APDRP) and the US–India Partnership to Advance Clean Energy-Development (PACE-D) programmes. Data management technologies and automatic screening of data, collected through remote terminal units (RTUs) is the worldwide challenge to make the network smart and to take quick decisions (ISGTF 2013). However, smart grid is a visionary and revolutionary change in the power sector which requires contributions from industry, academic, and research institutions. Smart grid architecture varies from place to place and essentially depends on the present grid structure, load dynamics, and resource availability. The Indian power sector still suffers from huge unmet demand due to lack of peak load management and high T & D losses. Smart grid would primarily reduce the network losses and narrow the energy demand gap. Power sector should be analysed, considering the future demand and then the grid architecture should be decided, whereas replicating the smart grid architecture may not be the solution.

4.4 Innovations in Energy Sector

Development of economically viable and technically feasible new energy harvesting technologies is expected to change the present energy mix. Technology innovation in non-fossil energy resources - solar thermal and PV, bioenergy, off-shore wind, hydrogen, artificial photosynthesis, etc. would meet the future energy demand (Abas et al. 2015). The current focus is on bioenergy, bio-oil, and biological hydrogen production. Technologies like bio-oil and ethanol production from algae would significantly replace the fossil oil for transportation and electricity generation (Gupta and Verma 2015). Many of these technologies are in the lab scale at the moment and thus, have shown great potential in cutting down the cost and also tapping a wide range of renewable energy sources.

Significant improvements are also found in energy storage technologies in order to resolve the intermittency issues in renewable energy sources. Table 1 summarizes a few energy storage technologies which are being developed and also used across the globe. However, industry collaboration would be necessary in order scale up/widen the new energy technologies and also for wide-scale dissemination.

In the face of increasing CO_2 emissions from conventional energy (gasoline) and the anticipated scarcity of crude oil, a worldwide effort is underway for cost effective renewable alternative energy sources. Efforts are in progress at Energy & Wetlands Research Group, CES (http://ces.iisc.ernet.in/energy), at the Indian Institute of Science, Banaglore, towards developing the gasoline secreting diatom solar panels to produce gasoline from diatoms sustainably. Diatoms being the major group of planktonic algae (Fig. 17) can be used sustainably for production of bio-fuel, by the usage of diatom-based solar panels. Studies have shown that diatoms could make 10 to 200 times as much oil per hectare as oil seeds (Ramachandra et al. 2009) and the techniques involved towards developing oil secreting diatoms to minimize the cost of oil extraction. It was found that some diatoms secrete more lipid content when subjected to unfavourable environment or culture conditions, such as nutrient starvation or extreme temperatures. Unlike crops, diatoms multiply rapidly. Some diatoms can double their biomass within an hour to a day's time. Since each diatom creates and uses its own gas tank, it is estimated that diatoms are responsible for up to 25% of global carbon dioxide fixation. This means that while diatoms can be cultivated for oil extraction, they can automatically reabsorb carbon dioxide in the process. Diatoms may have a major role to play in the coming years with regard to the mass production of oil. This entails appropriate cultivation, harvesting and extraction of oil, using advanced technologies that mimic the natural process while cutting down the time period involved in oil formation.

Energy from Wastes: Urban areas are generating a large quantum of waste. For example, Greater Bangalore generates about 1,200 MLD of liquid waste and about 2,800 tonnes of solid waste every day. Untreated wastes are contributing to greenhouse gases (GHG) in the system and also to global warming (Ramachandra 2009b). Viable technologies are available to convert waste to energy. For example, an algae photo-bioreactor that grows algae in municipal wastewater to produce

Table 1 Energy storage technologies (Decourt and Debarre 2013; Paksoy 2013)

Technology	Location	Output	Efficiency (%)	Initial investment cost (USD/kW)	Primary application
Pumped Storage	Supply	Electricity	50–85	500–4600	Long-term
Underground Thermal Energy Storage (UTES)	Supply	Thermal	50–90	3400–4500	Long-term storage
Compressed air storage	Supply	Electricity	27–70	500–1500	Long-term storage, arbitrage
Pit storage	Supply	Thermal	50–90	100–300	Medium temperature applications
Molten salts	Supply	Thermal	40–93	400–700	High-temperature applications
Batteries	Supply, demand	Electricity	75–95	300–3500	Distributed/off-grid storage, short-term storage
Thermochemical	Supply, demand	Thermal	80–99	1000–3000	Low, medium, and high-temperature applications
Chemical-hydrogen storage	Supply, demand	Electrical	22–50	500–750	Long-term storage
Flywheels	T&D	Electricity	90–95	130–500	Short-term storage
Supercapacitors	T&D	Electricity	90–95	130–515	Short-term storage
Superconducting magnetic energy storage (SMES)	T&D	Electricity	90–95	130–515	Short-term storage
Solid media storage	Demand	Thermal	50–90	500–3000	Medium temperature applications
Ice storage	Demand	Thermal	75–90	6000–15,000	Low-temperature applications
Hot water storage (residential)	Demand	Thermal	50–90	Negligible	Medium temperature applications
Cold-water storage	Demand	Thermal	50–90	300–600	Low-temperature applications

Source Abas et al. 2015

biofuel and a variety of other products is in place (Mahapatra et al. 2014). This bioreactor will not compete with agriculture for land, fertilizer, or freshwater. Similarly, to handle the organic fraction of municipal waste (which constitute 60–70% of Bangalore's municipal waste), Centre for Sustainable Technologies at the Indian Institute of Science (IISc), Bangalore, has developed a viable technology.

Fig. 17 Pennate and centric diatoms (*Navicula* sp., with an oil droplet)

The policy shift, political-will, and active participation of decision makers and all stakeholders (local community) are required to see these technologies are in place and Bangalore is free of wastes (Ramachandra 2009b; Chanakya et al. 2007a, b, 2009).

4.5 *Future Energy Scenario*

Natural resource exploitation in the country has increased manifold over the years to cater the energy requirements in all sectors. Resource extraction is forecasted till 2021 using the historical rate of consumption and is given in Fig. 18. It shows an increasing trend which necessitates the immediate energy conservation and

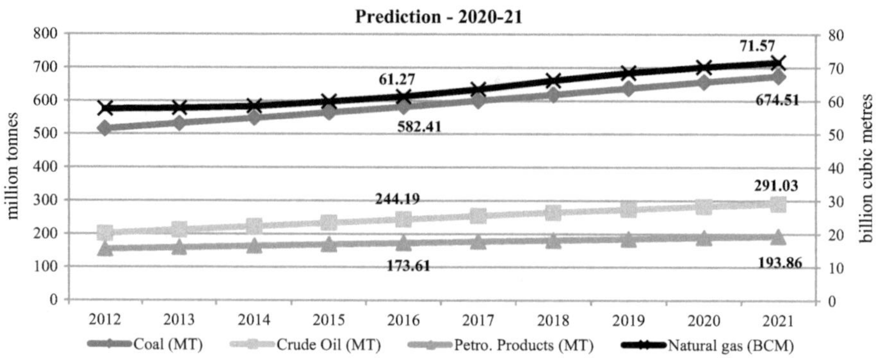

Fig. 18 Resource consumption prediction for 2021

exploitation of non-conventional sources of energy. Extraction of coal resources is projected as 674 million tonnes, which is about 34% more than the present consumption (2011). Estimation also reveals an increment of 38 and 31% in crude oil and petroleum production consumption. Increase in the natural gas consumption is expected to be marginal ($\sim$25%) with respect to the present consumption. This demands radical government policies focusing on renewable energy, revolutionary improvements in end-use technologies, and changes in the resource utilization practices. Nevertheless, the current trend of consumption of fossil fuel resources has caused many environmental problems, thus, necessitating restructuring of energy portfolio.

5 Conclusions and Recommendations

The Indian energy sector is at the tipping point as alternate renewable energy technologies have gained significance during the last two decades. There is a need to navigate the energy transition for sustainable growth in socio-economic aspects of the country. Though the energy consumption per GDP is higher, production of valuable goods is quite low in the country which shows that there is a need to improve the end-use efficiency. Energy utilization from non-conventional energy sources holds the major share after fossil fuels, which has to be considered for technology improvement. Indian electric power generation mainly depends on coal and hydro resources which are centralized. Sparsely located load centers, theft, pilferage, and unmetered supply causes high T & D losses which needs to be reduced through distributed generation and micro-grids. Capacity addition through renewable energy sources will ensure the effective renewable mix in total installed capacity, while meeting the future demand. Demand-side management and end use efficiency improvements are the short term requirements to fill the supply–demand gap and for reliable energy supply. It also reduces the derivatives (ash, fumes, GHG gases, etc.) by increasing the net productivity. Smart grid technology will make the power system more reliable, secure, efficient, and environment-friendly. New energy sources and effective use of available resources would keep the country on a sustainable path leading to achievement of energy independence.

The sustainable energy option requires the government support for the identification, exploitation and use of renewable sources of energy, which should be at least as high as for conventional sources. A generation based incentive (GBI) would encourage decentralized electricity generation at individual rooftops. In addition to this, there is a need to promote solar rooftops in Government infrastructure and buildings such as (i) solar powered street lights, (ii) install solar rooftops in all new government/local body buildings, (iii) implementation solar rooftops in a phased manner in the existing government/local body buildings, and solar power systems for all street lights and water supply installations in local bodies in a phased manner. The current study shows that the renewable energy experiences all the useful kinds of problems that affect most economic development projects. Lack of capital, skilled labour and service backups are serious impediments to the progress of alternate devices such as fuel efficient stoves, biogas, wood gasifiers, etc. Even more serious concern is the lack of coordinated effort among various bureaucratic setup and ministries. These are the main hurdles to the successful implementation of biomass cultivation projects and development of bioenergy. Policies are to be formulated to remove the constraints at local/regional level. Policies must engender the communication between the different institutions and government sectors involved with the establishment of a significant and sustainable bioenergy programme that is the agricultural, forestry, land planning and energy sectors. Hence, the prudent management practices involving energy generation from renewable sources, while meeting the energy requirements at decentralized levels efficiently would offer the opportunity to address multiple environmental concerns such as land degradation, bio diversity, acid rain pollutants, local and regional health problems.

Acknowledgements We are grateful to (i) the NRDMS Division, Ministry of Science and Technology (DST), Government of India; (ii) The Ministry of Environment, Forest and Climate Change (MoEFCC), Government of India; and (iii) Indian Institute of Science, Bangalore, for the sustained financial and infrastructure support.

References

Abas, N., Kalair, A., & Khan, N. (2015). Review of fossil fuels and future energy technologies. *Futures*, *69*, 31–49.

Albadi, M. H., & El-Saadany, E. F. (2007). Demand response in electricity markets: An overview. *Energy Policy*, *33*, 1021–1036.

Al-mulali, U., & Che, N. B. C. S. (2012). The impact of energy consumption and CO_2 emission on the economic growth and financial development in the Sub Saharan African countries. *Energy*, *39*, 180–186.

Asafu-Adjaye, J. (2000). The relationship between energy consumption, energy prices and economic growth: Time series evidence from Asian developing countries. *Energy Economics*, *22*, 615–625.

Balamurugan, P., Ashok, S., & Jose, T. L. (2009). Optimal operation of biomass/wind/PV hybrid energy system for rural areas. *International Journal of Green Energy*, *6*, 104–116.

Bhattacharyya, S. C. (2010). Shaping a sustainable energy future for India: Management challenges. *Energy Policy*, *38*, 4173–4185.

CEA. (2013). Annual Report 2012–13, Central Electricity Authority, Government of India. Retrieved September 14, 2013, from http://www.cea.nic.in.

CEA—LGBR, Load Generation Balance Report (LGBR). (2012–13) Central Electricity Authority, Government of India. Retrieved September 2013, from http://www.cea.nic.in.

C-WET. (2012). Annual Report 2011–12, Centre for Wind Energy Technology, Ministry of New and Renewable Energy, Government of India, Chennai, India.

Chanakya, H. N., Ramachandra, T. V., Guruprasad, M., & Devi, Vinutha. (2007a). Micro-treatment options for components of organic fraction of MSW in residential areas. *International Journal of Environmental Monitoring and Assessment*, *135*(1–3), 129–139.

Chanakya, H. N., Ramachandra, T. V., & Vijayachamundeeswari, M. (2007b). Resource recovery potential from secondary components of segregated municipal solid wastes. *International Journal of Environmental Monitoring and Assessment*, *135* (1–3), 119–127.

Chanakya, H. N., Sharma, Isha, & Ramachandra, T. V. (2009). Micro-scale anaerobic digestion of point source components of organic fraction of municipal solid waste. *Waste Management, 29* (4), 1306–1312.
Dabrase, P., & Ramachandra, T. V. (1999). Energy and environmental sustainability: Some key issues in rural Kolar, Karnataka, India. In *Criteria and Indicators of Sustainability in Rural Development—a Natural Resource Perspective*. New Delhi: Oxford and IBH Publishing Pvt. Ltd.
Decourt, B., & Debarre, R. (2013). *Electricity storage factbook*. Paris, France: Schlumberger Business Consulting Energy Institute.
EIA. (2013). Energy Information Administration 2013. Retrieved September 15, 2013, from http://www.eia.gov/countries/country-data.cfm?fips=IN&trk=m#tpe.
El-Khattam, W., Hegazy, Y. G., & Salama, M. M. A. (2005). An integrated distributed generation optimization model for distribution system planning. *IEEE Transactions on Power Systems, 20,* 1158–1165.
Energy Realities. (2013). Retrieved September 14, 2013, http://www.energyrealities.org/chapter/meeting-our-needs/item/per-capita-energy-consumption/erp327B7C729A3B31D2B.
Ghosh, D., Shukla, P. R., Garg, A., & Venkataramana, P. (2002). Renewable energy technologies for the Indian power sector: Mitigation potential and operational strategies. *Renewable and Sustainable Energy Reviews, 6,* 481–512.
Gupta, M., & Sengupta, R. (2012). Energy savings potential and policy for energy conservation in selected indian manufacturing industries. Working Paper No. 2012-105. New Delhi: National Institute of Public Finance and Policy.
Gupta, A., & Verma, J. P. (2015). Sustainable bio-ethanol production from agro-residues: A review. *Renewable and Sustainable Energy Reviews, 41,* 550–567.
Hiremath, R. B., Shikha, S., & Ravindranath, N. H. (2007). Decentralized energy planning; modeling and application—A Review. *Renewable and Sustainable Energy Reviews, 11,* 729–752.
ISGTF. (2013). Indian Smart Grid Task Force (ISGTF)—Smart Grid Pilot Projects under Execution in India (2013–2016), Ministry of Power, Government of India. Retrieved October 18, 2013, from http://www.isgtf.in/.
Kumar, A., Kumar, K., Naresh, K., Sharma, S., & Mishra, S. (2010). Renewable energy in India: Current status and future potentials. *Renewable and Sustainable Energy Reviews, 14,* 2434–2442.
Kumar, D. M., Christopher, S. A., & Singh, O. P. (2013). Can India raise agricultural productivity while reducing groundwater and energy use? *International Journal of Water Resources Development, 29,* 557–573.
Mahapatra, D. M., Chanakya, H. N., & Ramachandra, T. V. (2014). Bioremediation and lipid synthesis through mixotrophic algal consortia in municipal wastewater. *Bioresource Technology, 168,* 142–150.
MNRE. (2013a). Ministry of New and Renewable Energy, Government of India. Retrieved September 13, 2013 and October 12, 2015, from http://www.mnre.gov.in.
MNRE. (2013b). Ministry of New and Renewable Energy, National Biomass Cookstoves Programme. Retrieved October 10, 2013 and October 12, 2015, from http://mnre.gov.in/schemes/decentralized-systems/national-biomass-cookstoves-initiative/.
MoC. (2013). Ministry of Coal, Government of India. Retrieved September 13, 2013 and October 12, 2015, from http://www.coal.nic.in/welcome.html.
MoP. (2013). Annual Report 2012–13, Ministry of Power, Government of India. Retrieved September 13, 2013 and October 12, 2015, from http://powermin.nic.in/.
MoPNG. (2013). Indian Petroleum & Natural Gas Statistics 2011–12, Government of India, Ministry of Petroleum & Natural Gas, Government of India, New Delhi. Retrieved September 13, 2013 and October 12, 2015, from http://petroleum.nic.in/reports.htm.
Nouni, M. R., Mullick, S. C., & Kandpal, T. C. (2008). Providing electricity access to remote areas in India: An approach towards identifying potential areas for decentralized electricity supply. *Renewable and Sustainable Energy Reviews*, *12*, 1187–1220.

Nouni, M. R., Mullick, S. C., & Kandpal, T. C. (2009). Providing electricity access to remote areas in India: Niche areas for decentralized electricity supply. *Renewable Energy, 34,* 430–434.

NSSO. (2007). Energy sources for Indian households for cooking and lighting 2004–05. 2007. NSS 61st round. National Sample Survey Organization (NSSO), Ministry of Statistics and Programme Implementation, Government of India.

Pachauri, S., & Jiang, L. (2008). The household energy transition in India and China. *Energy Policy, 36,* 4022–4035.

Paksoy, H. (2013). Thermal energy storage today. Paper Presented at the IEA Energy Storage Technology Roadmap Stakeholder Engagement Workshop. Paris, France.

Palensky, P., & Dietrich, D. (2011). Demand side management: Demand response, intelligent energy systems, and smart loads. *IEEE Transactions on Industrial Informatics, 7,* 381–388.

Pathomthat, C., & Ramakumar, R. (2004). An approach to quantify the technical benefits of distributed generation. *IEEE Transactions on Energy Conversion, 19,* 764–773.

Prakash, V. R., & Bhat, I. K. (2009). Energy, economics and environmental impacts of renewable energy systems. *Renewable and Sustainable Energy Reviews, 13,* 2716–2721.

Ramachandra, T. V. (2008). Geographical information system approach for regional biogas potential assessment. *Research Journal of Environmental Sciences, 2,* 170–184.

Ramachandra, T. V. (2009a). RIEP: Regional Integrated Energy Plan. *Renewable and Sustainable Energy Reviews, 13,* 285–317.

Ramachandra, T. V. (2009b). *Management of municipal solid waste*. New Delhi: TERI Press.

Ramachandra, T. V. (2011). Renewable energy transition: Perspective and Challenges. In *Energy India 2020—A Shape of Things to come in Indian Energy Sector* (pp. 175–183). Ahmedabad: Saket Projects Ltd.

Ramachandra, T. V., & Krishnadas, G. (2012). Prospects and Challenges of decentralized wind applications in the Himalayan Terrain. *Journal of Energy Bioscience, 3,* 1–12.

Ramachandra, T. V., & Shruthi, B. V. (2005). Wind energy potential mapping in Karnataka, India using GIS. *Energy Conversion and Management, 46,* 1561–1578.

Ramachandra, T. V., & Shruthi, B. V. (2007). Spatial mapping of renewable energy potential. *Renewable and Sustainable Energy Reviews, 11*, 1460–1480.

Ramachandra, T. V., & Subramanian, D. K. (1995). Industrial energy utilisation in Karnataka and potential savings. *Energy Conversion and Management, 38,* 563–599.

Ramachandra, T. V., Subramanian, D. K., Joshi, N. V., Gunaga, S. V., & Harikantra, R. B. (1999). End use efficiencies in the domestic sector of Uttara Kannada District. *Energy Conversion and Management, 41*, 833–845.

Ramachandra, T. V., Subramanian, D. K., Joshi, N. V., Gunaga, S. V., & Harikantra, R. B. (2000). End use efficiencies in the domestic sector of Uttara Kannada District. *Energy Conversion and Management, 41*, 833–845.

Ramachandra, T. V., Loerincik, Y., & Shruthi, B. V. (2006). Intra and inter country energy intensity trends. *International Journal of Energy and Development, 31,* 43–84.

Ramachandra, T. V., Mahapatra, D. M., Karthick, B., & Gordon, R. (2009). Milking diatoms for sustainable energy: Biochemical engineering versus gasoline secreting diatom solar panels. *Industrial and Engineering Chemistry Research, 48,* 8769–8788.

Ramachandra, T. V., Jain, R., & Krishnadas, G. (2011). Hotspots of solar potential in India. *Renewable and Sustainable Energy Reviews, 15*, 3178–3186.

Ramachandra, T. V., Ramakrishna, Y. B., Krishnadas, G., Sudarshan Bhat, P., Mahapatra, D. M., & Bharath Aithal, H. (2012). Environmental profile and people's livelihood aspects in the vicinity of coal based thermal power plant at Yellur Panchayat, Udupi District. CES Technical Report: 126, Energy & Wetlands Research Group, Centre for Ecological Sciences. Bangalore: Indian Institute of Science.

Ramachandra, T. V., Madhab, D. M., Samantray, S., & Joshi, N. V. (2013). Algal biofuel from urban wastewater in India: Scope and challenges. *Renewable and Sustainable Energy Reviews, 21,* 767–777.

Ramachandra, T. V., Hegde, G., Setturu, B., & Krishnadas, G. (2014a). Bioenergy: A sustainable energy option for rural India. *Advances in Forestry Letters (AFL), 3,* 1–15.

Ramachandra, T. V., Hegde, G., & Krishnadas, G. (2014b). Scope of solar energy in Uttara Kannada, Karnataka State, India: Roof top PV for domestic electricity and standalone systems for irrigation. *Productivity, 55,* 100–119.

Ramachandra, T. V., Hegde, G., & Krishnadas, G. (2014c). Potential assessment and decentralized applications of wind energy in Uttara Kannada, Karnataka. *International Journal of Renewable Energy Research, 4,* 1–10.

Ravindranath, N. H., & Balachandra, P. (2009). Sustainable bioenergy for India: Technical, economic and policy analysis. *Energy, 34,* 1003–1013.

Reddy, A. K. N. (1999). Goals, strategies and policies for rural energy. *Economic and Political Weekly, 34,* 3435–3445.

Rehman, I. H., Khar, A., Banerjee, M., Kumar, P., Shardul, M., Mohanty, J., & Hossain, I. (2012). Understanding the political economy and key drivers of energy access in addressing national energy access priorities and policies. *Energy Policy*, *47*, 27–37.

Sharma, A., Srivastava, J., Kar, S., & Kumar, A. (2012). Wind energy status in India: A short review. *Renewable and Sustainable Energy Reviews, 16,* 1157–1164.

Simron, J. S., Krausmann, F., Gingrich, S., Haberl, H., Erb, K. H., Lanz, P., et al. (2012). India's biophysical economy 1961–2008: Sustainability in a national and global context. *Ecological Economics, 76,* 60–69.

Singh, R., & Setiawan, A. D. (2013). Biomass energy policies and strategies: Harvesting potential in India and Indonesia. *Renewable and Sustainable Energy Reviews, 22,* 332–345.

TEDDY. (2013). *TERI Energy Data Directory and Year Book (TEDDY).* The Energy and Resources Institute (TERI): New Delhi.

Vijayapriya, T., & Kothari, D. P. (2011). Smart grid: An overview. *Smart Grid and Renewable Energy, 2,* 305–311.

Vishwanathan, B., & Ravikumar, K. S. (2005). Cooking fuel use patterns in India: 1983–2000. In *IEEE Power Engineering Society General Meeting*, 24th–28th June 2007, Tampa, Florida, USA.